THOMAS MINCHIN GOODEVE
CHARLES PERCY BYSSHE SHELLEY

THE WHITWORTH MEASURING MACHINE

Elibron Classics
www.elibron.com

Elibron Classics series.

ISBN 1-4021-9817-5 (paperback)
ISBN 1-4021-0910-5 (hardcover)

This Elibron Classics Replica Edition is an unabridged facsimile of the edition published in 1877 by Longmans, Green, and Co., London.

TEXT-BOOKS OF SCIENCE.

In course of publication, in small 8vo. each volume complete in itself,

A SERIES OF

ELEMENTARY WORKS ON MECHANICAL AND PHYSICAL SCIENCE,

FORMING A SERIES OF

TEXT-BOOKS OF SCIENCE

ADAPTED FOR THE USE OF ARTISANS AND OF STUDENTS IN PUBLIC AND SCIENCE SCHOOLS.

EIGHT TEXT-BOOKS PREPARING FOR PUBLICATION.

PHOTOGRAPHY.
By Captain ABNEY, Royal Engineers, F.R.S. Instructor in Chemistry and Photography at the School of Military Engineering, Chatham.

THEORY OF SOUND.
By W. G. ADAMS, M.A. Professor of Natural Philosophy and Astronomy, King's College, London.

ASTRONOMY.
By ROBERT S. BALL, LL.D. F.R.S. Andrews Professor of Astronomy in the University of Dublin, and Royal Astronomer of Ireland.

MINERALOGY.
By H. BAUERMAN, F.G.S. Associate of the Royal School of Mines.

PHYSICAL GEOGRAPHY.
By the Rev. GEORGE BUTLER, M.A. Principal of Liverpool College; Editor of 'The Public Schools Atlas of Modern Geography.'

ECONOMICAL APPLICATIONS OF HEAT.
Including Combustion, Evaporation, Furnaces, Flues, and Boilers. By C. P. B. SHELLEY, Civil Engineer, and Professor of Manufacturing Art and Machinery at King's College, London. With a Chapter on the Probable Future Development of the Science of Heat by C. WILLIAM SIEMEN, F.R.S.

PHYSICAL OPTICS.
By G. G. STOKES, M.A. D.C.L. Fellow of Pembroke College, Cambridge; Lucasian Professor of Mathematics in the University of Cambridge; and Secretary to the Royal Society.

AGRICULTURAL CHEMISTRY.
By ROBERT WARRINGTON, F.C.S.

TWENTY TEXT-BOOKS NOW PUBLISHED.

THE ELEMENTS OF MECHANISM.
Designed for Students of Applied Mechanics. By T. M. GOODEVE, M.A. Barrister-at-Law, Lecturer on Mechanics at the Royal School of Mines. New Edition, revised; with 257 Figures on Wood. Price 3*s.* 6*d.*

METALS, THEIR PROPERTIES AND TREATMENT.
By CHARLES LOUDON BLOXAM, Professor of Chemistry in King's College, London; Professor of Chemistry in the Department of Artillery Studies, and in the Royal Military Academy, Woolwich. With 105 Figures on Wood. Price 3*s.* 6*d.*

INTRODUCTION TO THE STUDY OF INORGANIC CHEMISTRY.
By WILLIAM ALLEN MILLER, M.D. LL.D. F.R.S. late Professor of Chemistry in King's College, London: Author of 'Elements of Chemistry, Theoretical and Practical.' New Edition, revised; with 71 Figures on Wood. Price 3*s.* 6*d.*

ALGEBRA AND TRIGONOMETRY.
By the Rev. WILLIAM NATHANIEL GRIFFIN, B.D. sometime Fellow of St. John's College, Cambridge. Price 3*s.* 6*d.*

NOTES ON THE ELEMENTS OF ALGEBRA AND TRIGONOMETRY;
With SOLUTIONS of the more difficult QUESTIONS. By the Rev. WILLIAM NATHANIEL GRIFFIN, B.D. sometime Fellow of St. John's College, Cambridge. Price 3*s.* 6*d.*

PLANE AND SOLID GEOMETRY.

By the Rev. H. W. WATSON, formerly Fellow of Trinity College, Cambridge, and late Assistant-Master of Harrow School. Price 3*s.* 6*d.*

THEORY OF HEAT.

By J. CLERK MAXWELL, M.A. LL.D. Edin. F.R.SS. L. & E. Professor of Experimental Physics in the University of Cambridge. New Edition, revised; with 41 Woodcuts and Diagrams. Price 3*s.* 6*d.*

TECHNICAL ARITHMETIC AND MENSURATION.

By CHARLES W. MERRIFIELD, F.R.S. an Examiner in the Department of Public Education, and late Principal of the Royal School of Naval Architecture and Marine Engineering, South Kensington. Price 3*s.* 6*d.*

KEY TO MERRIFIELD'S TEXT-BOOK OF TECHNICAL ARITHMETIC AND MENSURATION.

By the Rev. JOHN HUNTER, M.A. one of the National Society's Examiners of Middle-Class Schools, formerly Vice-Principal of the National Society's Training College, Battersea. Price 3*s.* 6*d.*

ON THE STRENGTH OF MATERIALS AND STRUCTURES.

The Strength of Materials as depending on their quality and as ascertained by Testing Apparatus; the Strength of Structures as depending on their form and arrangement, and on the materials of which they are composed. By JOHN ANDERSON, C.E. LL.D. F.R.S.E. Superintendent of Machinery to the War Department. Price 3*s.* 6*d.*

ELECTRICITY AND MAGNETISM.

By FLEEMING JENKIN, F.R.SS. L. & E. Professor of Engineering in the University of Edinburgh. New Edition, revised. Price 3*s.* 6*d.*

WORKSHOP APPLIANCES.

Including Descriptions of the Gauging and Measuring Instruments, the Hand Cutting Tools, Lathes, Drilling, Planing, and other Machine Tools used by Engineers. By C. P. B. SHELLEY, Civil Engineer, Hon. Fellow and Professor of Manufacturing Art and Machinery at King's College, London. With 209 Figures on Wood. Price 3*s.* 6*d.*

PRINCIPLES OF MECHANICS.

By T. M. GOODEVE, M.A. Barrister-at-Law, Lecturer on Applied Mechanics at the Royal School of Mines. Price 3*s.* 6*d.*

INTRODUCTION TO THE STUDY OF ORGANIC CHEMISTRY.

The CHEMISTRY of CARBON and its COMPOUNDS. By HENRY E. ARMSTRONG, Ph.D. F.C.S. Professor of Chemistry in the London Institution. With 8 Figures on Wood. Price 3*s.* 6*d.*

MANUAL OF QUALITATIVE ANALYSIS AND LABORATORY PRACTICE.

By T. E. THORPE, Ph.D. F.R.S.E. Professor of Chemistry in the Andersonian University, Glasgow; and M. M. PATTISON MUIR. Price 3*s.* 6*d.*

QUANTITATIVE CHEMICAL ANALYSIS.

By T. E. THORPE, F.R.S.E. Ph.D. Professor of Chemistry in the Andersonian University, Glasgow. With 88 Figures on Wood. Price 3*s.* 6*d.*

TELEGRAPHY.

By W. H. PREECE, C.E. Divisional Engineer, Post-Office Telegraphs; and J. SIVEWRIGHT, M.A. Superintendent (Engineering Department) Post-Office Telegraphs. With 160 Figures and Diagrams. Price 3*s.* 6*d.*

RAILWAY APPLIANCES.

A Description of Details of Railway Construction subsequent to the completion of the Earthworks and Masonry; including a short Notice of Railway Rolling Stock. By J. W. BARRY, Member of the Institution of Civil Engineers. With 207 Figures and Diagrams. Price 3*s.* 6*d.*

INTRODUCTION TO THE STUDY OF CHEMICAL PHILOSOPHY; THE PRINCIPLES OF THEORETICAL AND SYSTEMATIC CHEMISTRY.

By WILLIAM A. TILDEN, B.Sc. London, F.C.S. Lecturer on Chemistry in Clifton College. With 5 Figures engraved on Wood. Price 3*s.* 6*d.*

STRUCTURAL AND PHYSIOLOGICAL BOTANY.

By Dr. OTTO WILHELM THOMÉ, Ordinary Professor of Botany at the School of Science and Art, Cologne. Translated and Edited by A. W. BENNETT, M.A. B.Sc. F.L.S. Lecturer on Botany at St. Thomas's Hospital. With 600 Woodcuts and a Coloured Map. Price 6*s.*

ELEMENTS OF MACHINE DESIGN:

An Introduction to the Principles which determine the Arrangement and Proportions of the Parts of Machines and a Collection of Rules for Machine Design. By W. CAWTHORNE UNWIN, B.Sc. Assoc. Inst. C.E. Professor of Hydraulic and Mechanical Engineering at the Royal Indian Civil Engineering College, Cooper's Hill. With 219 Wood Engravings. Price 3*s.* 6*d.*

THE ART OF ELECTRO-METALLURGY,

Including all Known Processes of Electro-Deposition. By G. GORE, F.R.S. With 56 Wood Engravings. Price 6*s.*

THE

WHITWORTH MEASURING MACHINE

LONDON: PRINTED BY
SPOTTISWOODE AND CO., NEW-STREET SQUARE
AND PARLIAMENT STREET

THE

WHITWORTH MEASURING MACHINE

INCLUDING DESCRIPTIONS OF THE SURFACE PLATES,
GAUGES, AND OTHER MEASURING INSTRUMENTS, MADE BY
SIR JOSEPH WHITWORTH, BART., C.E. F.R.S. D.C.L. LL.D. &c.

BY

T. M. GOODEVE, M.A.

Barrister-at-Law: Lecturer on Applied Mechanics at the Royal School of Mines

AND

C. P. B. SHELLEY

Civil Engineer: Honorary Fellow of, and Professor of Manufacturing Art and Machinery in, King's College, London

WITH 4 PLATES AND 44 WOODCUTS

LONDON
LONGMANS, GREEN, AND CO.
1877

LIST OF ILLUSTRATIONS.

THE WHITWORTH
MEASURING MACHINE.

CHAPTER I.

IT IS WELL KNOWN that the mechanician who is engaged in constructing machinery of precision relies entirely upon the power of producing true surfaces. The efficiency of a machine, when completed, may depend so largely on the truth of the surfaces of the moving parts, that it becomes of the highest importance to eliminate as far as may be possible any errors which might detract from that extreme accuracy of movement which alone can produce successful results.

The two principal surfaces of essential importance in the workshop may be distinguished as a 'true plane,' and a 'true cylinder;' and it is necessary, at the outset of our enquiries, to attach some definite meaning to these expressions. A 'true plane' is manifestly an ideal conception which, like any other perfect thing, may be striven for, but cannot be attained. It is, of course, not within the limits of constructive power to make a metallic plane surface which is absolutely true in any sense. The nearest approach to it is probably the surface of clean mercury when at rest; and the high reflective power of the metal enables us to observe, when looking on a bowl of mercury, the formation of brilliant images of external objects due to the truth of a plane formed in obedience to the laws of fluid pressure.

It has thus become a common thing to regard the polish of any artificial surface as a test of its truth.

But the idea of the necessity of polish must be abandoned altogether when considering a metallic plane surface as aimed at and constructed for the use of the mechanician. The approximate true plane which is of so much value in the workshop has no polish at all; it is mottled throughout by the action of the scraping tool which has operated on it, and is to be regarded as 'true,' not in the optical sense of being a correct reflector, but in the mechanical sense of supplying a vast assemblage of minute bearing surfaces evenly distributed and approaching closely to one geometrical bounding plane. Such an approximate plane when formed in iron and coated with a film of silver may receive a high degree of polish, and will then approach the truth of the surface of still mercury, with the advantage of being capable of being supported at an inclination to the horizon. The astronomers of the next generation may possibly appreciate the value of this extended use of the mechanician's primary surface.

At the present time the use of 'true planes' (called surface plates) is confined to the workshop, where such surfaces are invaluable, as being concerned either directly or indirectly in the production of all other surfaces of precision.

The conditions which a surface plate should fulfil are the following:—1. The bearing faces should all lie in the same plane. 2. They should be distributed as nearly as possible at equal distances from each other. 3. They should be sufficiently numerous for the particular application intended.

As to the first condition, it is easy to make the statement, but it is obvious that we have no power of ascertaining to what extent it is fulfilled, or how nearly it is approached. The degree of approximation to absolute truth must ever remain a matter of conjecture. As to the second and third conditions, we have them under control. An examination of the surface will at once reveal the bearing faces, and they may be crowded together more or less closely according to the will of the operator. In the case of sliding surfaces, the bearing faces should be evenly distributed, in order that the wear may be uniform and regular; and they should be closely arranged, in order that each part of the surface may duly support its portion of the work. But where two surfaces remain in contact without motion, as in the union of parts of the framework of a machine, the general approach to truth of position is the important consideration, and the bearing faces may be few in number and disposed at wide intervals, provided only that they all approach closely to one bounding plane.

It is easy to point to examples in illustration of these remarks. A planing machine is an instrument, as its very name implies, for multiplying plane surfaces, and accordingly there are 'mechanical true planes' (as we may agree to call them) carefully placed on each side of the machine, which guide the path of a table carrying the piece of metal operated on, and transfer the truth of their own surfaces to the work done as nearly as the imperfections of the cutting tool will permit. The slide rest of a lathe is built up of plane surfaces sliding upon each other. The bed of a lathe must be a plane, or it would be inadequate for the direction of the slide rest. The lathe and the planing machine are continually producing surfaces required for the construction of other machine tools, and it is therefore most necessary that they should themselves be perfect models. It is indeed obvious that any defects or errors of construction which exist in the leading machine will influence the whole work done, and may propagate such deviations from truth in an aggravated form. And not only so, but, looking beyond the workshop, it will be found impossible to separate our successful manufacturing skill from that primary element on which it is founded, namely, the mechanical power of producing true geometrical forms. For this reason it has been well said 'that a true plane is the foundation and source of all truth in mechanism.'

A remarkable advance in the direction to which we have referred was made prior to the year 1840 by Sir Joseph Whitworth, and formed the subject-matter of a paper read before the meeting of the British Association at Glasgow in that year. Specimens of truly plane metallic surfaces were then, for the first time, brought under the notice of scientific men, and the method of preparing such surfaces was also made known. Up to that time the process relied upon for obtaining plane surfaces on metal plates, and indeed the only one practically used, had involved the operation of grinding two plates together with emery powder and water. The plates, when under preparation, were occasionally compared with the surface of a standard plate, and the inequalities were removed. But there was no reliable mode of constructing the standard plate, the method of comparison was uncertain or imperfect, and the operation of grinding was deemed essential for bringing the plate to its most highly finished condition. In other words, the process was carried on under a mistaken conception of what ought to have been done, and without any knowledge that the problem of making a true plane could never be solved by the operation of grinding alone, for the simple reason that the action of grinding powder could not be restricted to those parts only which were in error. The grinding powder would necessarily diffuse itself unequally between two plates in contact, and the attempt to correct one class of errors would assuredly give rise to another which had not previously existed.

The object being to obtain a surface in which an assemblage of minute bearing surfaces (in reality almost points) are distributed evenly upon an imaginary geometrical plane, it is clear that a complete control over the successive removals of any portions of the surface believed to be in error is the first thing to be sought for. It is in this particular that the process of grinding fails, and the smooth surface given by grinding is entirely deceptive when accepted as an evidence of truth. The problem before us may be simple in character and may appear unimportant, but it is not so really; its solution has produced results of extraordinary value in advancing our power of construction, and has moreover led to a refinement of measurement which far exceeds anything that would have been deemed possible a few years ago. It is no exaggeration to say that linear intervals may now be measured mechanically which are too minute to be seen or recognised by the aid of any microscope.

Before entering upon the subject of measurement we propose to examine the method of preparing plane metallic surfaces in the manner made known by Sir J. Whitworth at the Glasgow meeting.

The surface plate, as then exhibited, was made of cast iron, and consisted of a rectangular plane table, ribbed at the back, and resting upon three bearing faces.

The form to be assigned to the plate is material when the question of lifting or supporting it comes under consideration. The conclusion, based on elementary propositions, may be stated in a very few words, viz.:—That where a plate is required to press equally on its three supports, or to be lifted by equal tensions, it must be constructed in such a manner that its centre of gravity shall coincide with that of the triangle formed by joining the points of support or of suspension.

In order to give effect to this statement an hexagonal or triangular form may be conveniently selected for the configuration of the surface, the points of support being

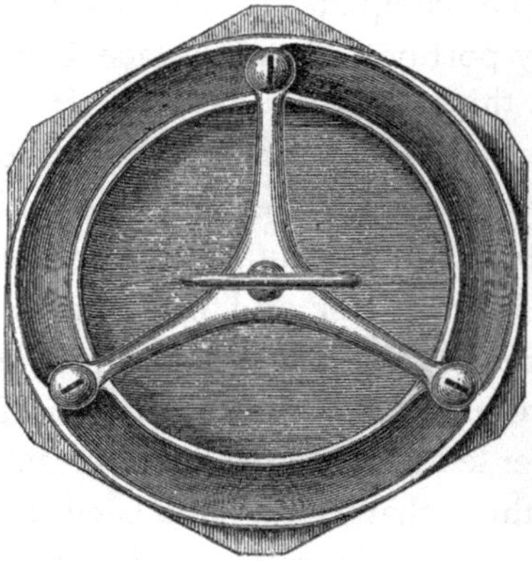

FIG. 1.

FIG. 2.

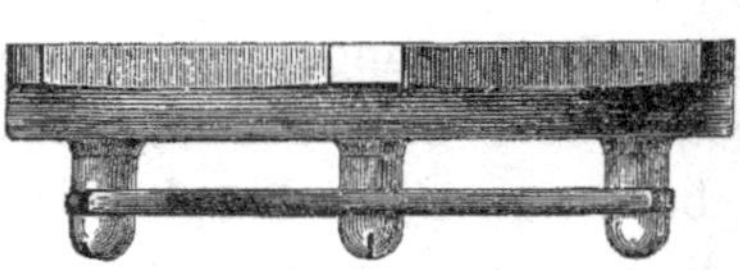

FIG. 3.

Plan and Elevations of Hexagonal Surface-plate.

symmetrically disposed at the angles of an equilateral triangle. The centre of figure of the hexagon or of the triangle will coincide with the centre of gravity of the space enclosed between the legs. And in respect of the ribs, it will be imperative so to arrange them in a symmetrical form that the common centre of gravity of the ribs and the plate shall not be thrown out of its first position as given by the form of the plate itself.

Sir Joseph Whitworth has lately patented an hexagonal surface plate (see figs. 1, 2, and 3), with the view of preventing irregular straining, the points of attachment employed for lifting being identical with the points of support.

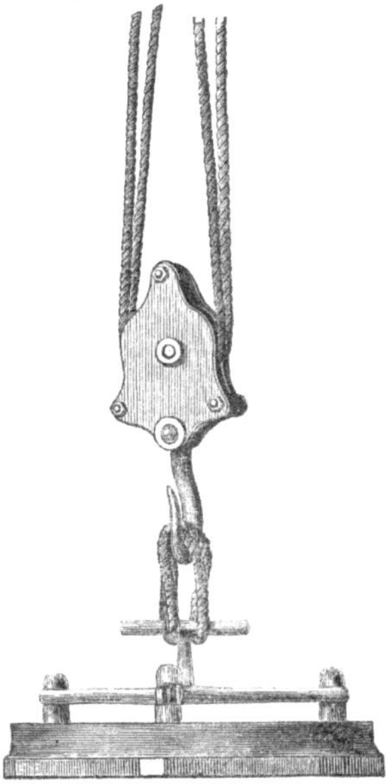

FIG. 4. Surface-plate suspended.

This construction reduces the amount of distortion caused by unequal strains to the smallest practicable amount, and further ensures that the plate, when suspended, is under the same conditions as if it were resting in the usual manner on a bench.

The points of support of an hexagonal surface plate are connected by a tripod frame, as shown in fig. 1, and the plate is slung from the central point of the frame as shown in fig. 4.

The table is therefore under the same strains as if it were swung independently at three points.

Surface plates made on the principle now referred to are found to possess so many advantages over the old rectangular plates, that Sir Joseph Whitworth has adopted them throughout his workshops, and has ceased to use any others for general purposes.

We may here remark that considerable interest has lately been exhibited in relation to experiments made with a view of testing the nature and truth of the surface formed upon hexagonal plates.

If two well-finished surface-plates be wiped with a dry cloth and laid one upon the other, the upper plate will appear to float and become buoyant, as if some lubricating matter existed between them; and the effect has been attributed to the presence of a film of air between the surfaces, which (it was said) relieved the friction so completely that they would move with a touch. If the upper plate be slightly raised and allowed to fall, there will be no metallic ring, but the blow emits a peculiar muffled sound, due (as believed) to the presence of a cushion of air. If a film of gold-leaf be placed between the plates, every atom of it disappears when the surfaces are rubbed together.

Again, if one surface be carefully slid on the other, so as to exclude the air, the plates adhere together with considerable force.

Dr. Tyndall remarked, in a lecture at the Royal Institution (June 4, 1875), that this effect was formerly described and accounted for as follows:—'If one plane be carefully slid over the other so as to exclude the air, the plates will adhere together with considerable force by the pressure of the atmosphere.' 'This clinging together of flat surfaces had been noticed before the time of Robert Boyle, and that great experimenter and noble philosopher first gave currency to the explanation here adopted. Boyle experimented with slabs of marble and with plates of glass, and thought he had proved that the clinging together did not occur in vacuo.' Dr. Tyndall here tested the result experimentally, and showed that 'two exceedingly accurate hexagonal Whitworth planes remained adherent in the best vacuum obtainable by a good air-pump. The vacuum was still further improved by filling the receiver with carbonic acid, and absorbing the residue with caustic potash. In this way the atmosphere was reduced until its total pressure on the surface of the hexagon amounted to only half a pound. The lower plate weighed 3 lbs., and to it was attached a mass of lead weighing 12 lbs. Though the pull of gravity was here thirty times the pressure of the atmosphere, the weight was supported. Indeed, it was obvious when an attempt was made to pull the plates asunder, that had a weight of 100 lbs. instead of 12 lbs. been attached to the lower hexagon, it would also have been sustained by the powerful attraction of the two surfaces.

'To show the probable character of the contact between the planes, two very perfect surfaces of glass were squeezed together with sliding pressure. They clung apparently as firmly as the Whitworth planes. Throwing by reflection from the glass plates a beam of light upon a white screen, the colours of "their plates" were vividly revealed. Clasping the plates of glass by callipers and squeezing them, the colours passed through various changes. When monochromatic light was employed, the successions of light and dark were numerous and varied, producing patterns of great beauty. All this proved that, though in such close mechanical contact, the plates were by no means in optical contact, being separated by distances capable of embracing several wave-lengths of the monochromatic light.'

It is also noticed that the amount of vacuum formed in

the receiver of an ordinary air-pump has little or no power in diminishing the floating effect which is observed when one plate lies upon the other. The floating is just as apparent under the exhausted receiver as in the open air. The explanation of this fact is not very obvious, but it is generally considered that the adhesion is caused by molecular attraction.

We pass on to describe the operation of preparing a standard surface plate, and shall select the hexagonal form by way of example.

FIG. 5. Chuck for attaching Plates to Table of Planing Machine.

A number of sound castings are obtained in the first instance, and the plates are fixed upon the table of a good planing machine. The faces of the bosses, being the bearing surfaces on which the plates are to be supported, will then be reduced to one level; and the bosses themselves are afterwards bored and tapped, in order that the screwed holes may afford a ready means of attaching the plates to the table of the planing machine. For the attachment chucks may be employed, a bolt being screwed upwards into each boss, and the chucks being secured in the ordinary way to the table of the machine. Fig. 5 shows one of the chucks employed. The plates are thus held firmly and securely, without any danger of being sprung or distorted by the unequal grip of clamps at the sides. They are then carefully planed over, and allowed to rest for two or three weeks, so that every portion of the plate may have time to assume its natural and unrestrained position. The workman now selects three plates out of the number and is ready to commence the task of converting their faces into mechanical true planes.

The first step is to ascertain which is the most perfect plate of the three chosen, and this may be done by laying a good straight-edge in various directions upon each surface and noting that which agrees most approximately with it. [A straight-edge is a flat steel bar on the thin edge of which a plane surface has been formed. In practice the surface of a straight-edge is a true plane, but for our present purpose it will be sufficient to consider it as an approximate true plane.] The most perfect plate of the three is regarded as the primary model, and the correction by the operation of scraping is commenced.

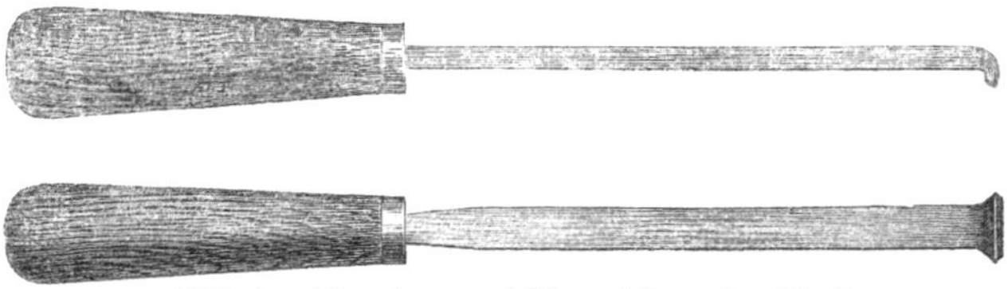

FIG. 6. Elevation and Plan of Scraping Tool.

A scraping tool, as shown in elevation and plan, fig. 6, is forged from a bar of steel, the size of the bar being dependent on the required width of the scraping edge. The bar is drawn out so as to be somewhat widened and made thinner at the end, where it is turned down and tempered, a cutting edge being formed by grinding two facets at an angle of about 60° to each other as shown at (A) in fig. 7. The edge is then carefully set on an oilstone, an operation which is frequently repeated when the tool is in use. As the scraping progresses, the angle of the cutting edge should be less acute, and is increased to a right angle, as shown at (B) in the intermediate stage; the angle being further increased to 120°, as shown at (C), for the finishing touches. The width

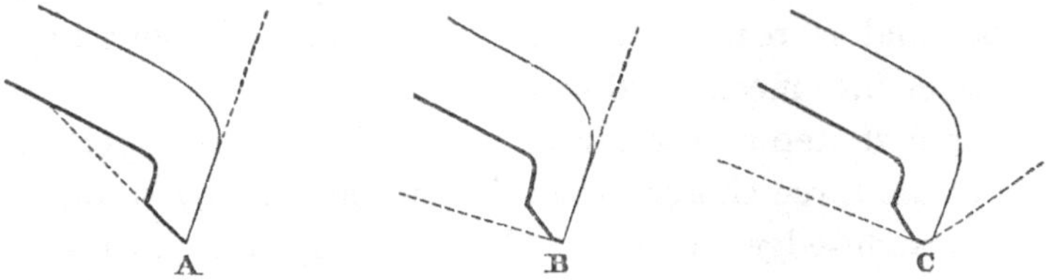

FIG. 7. Angles of Edges on Scraping Tools.

of the scraping edge varies from two and a half inches to half an inch, the workman using a broader tool in the preliminary stages, and reserving the narrower and lighter one for finishing. The form of the scraper may be varied, and worn-out flat files answer very well for conversion into

FIG. 8. Method of holding Scraping Tool.

scraping tools, the cutting edge being formed on the end of the file and at right angles to its length.

The method of holding the scraper will be apparent from fig. 8. The workman takes the tool in his right hand, pressing its edge upon the surface to be scraped with his left hand, and at the same time moves the tool to and fro over a small space, thus taking off very small quantities of metal in the form either of minute shavings or fine powder, according to the degree of force exerted.

It will be understood that scraping is a very delicate application of the principle of filing, the edge of the scraper acting as if it were one tooth of a very fine file; the point aimed at being to detach from any portion of a surface as much or as little of the metal as may be desired, and especially to confine the operation to any particular part at the precise spot which may be in error.

FIG. 9. Appearance of Scraped Surface.

In order to present some idea of the result of scraping on the appearance of a surface, the annexed drawing (fig. 9) has been engraved from a photograph, and exhibits the peculiar mottled character which is due to the multiplied action of the scraping edge in every direction as the work advances.

The operation of scraping having been described, we proceed to call it into our service in order to obtain a perfect coincidence between three surfaces, A, B, and C. It will be apparent that if the surfaces B and C coincide exactly with each other as well as with A, they must be all truly plane. This is a geometrical consequence which follows from the nature of a plane surface, and is merely an extension of that which is manifestly true in the case of lines. In other words, if three surfaces, A, B, C, can be brought to exact coincidence when compared and interchanged, each of them must be truly plane.

Taking two of the surfaces, as A and B, we have now to bring them into perfect coincidence. The face of A is very lightly smeared with red ochre and oil in an even film, and B is lowered upon A, and moved slightly, so that the colouring matter may adhere to the surface of B wherever there is contact. B is then scraped at all those points to which the colouring matter has adhered, A in its turn being wiped clean, and B being coated with the mixture, after which the operation is reversed and A is submitted to the scraper. The process is continued until the contact between A and B is made as perfect as possible, the bearing points being so numerous and evenly distributed that the entire surface appears reddened.

In order to secure the equal advance of all the parts together, particular attention must be paid to the colouring matter, with reference both to the quantity employed and to its equal distribution. If too small a quantity be used in the first instance, it will afford no evidence of the general condition of the surface. It will merely indicate the particular points which happen to be most prominent, and to reduce these in detail would be only a waste of time, so long as they are considerably above the general level.

When the finished surface is rubbed on a standard plate, the bearing points will become bright, and the operator will be able to judge of the degree of accuracy to which it has been brought. If it be as nearly true as it can be made by the hand, bright points will be seen diffused throughout its whole extent, interspersed with others that are less luminous, indicating thereby the degree of pressure which they respectively bear when in contact.

But the work is not yet completed, for although A and B are in perfect contact throughout, we are not justified in concluding that the surfaces are plane; we can only say that if one, such as B, be convex in any part, the other—viz. A—is concave in a corresponding degree, as shown in fig. 10.

But if we now take a third surface, such as C, and bring that into perfect coincidence with B by scraping, fig. 11, and then compare A and C together (as in fig. 12) the error, if

FIG. 10.

FIG. 11.

FIG. 12.

FIG. 13.

any, will be made manifest by the want of absolute contact, as indicated by the colouring matter. To bring A and C nearer the truth equal quantities must be scraped away from both surfaces at the points in contact. When this has been done with all the skill the mechanic may possess, and A and C are brought into coincidence with each other as in fig. 13, the next step is to bring up B to both, applying B to A and C in immediate succession, so as to compare the impressions. The art here lies in getting B between A and C in the probable direction of the true plane. Taking then the plate B as an improved standard for another comparison with A and C, the process commences *de novo*, and is carried on in a regular series of comparisons, which result in a gradual approach towards absolute truth. At last the inherent imperfections of the material and of the tools render it impossible to proceed further, and the most watchful care is necessary in order to guard against the introduction of fresh errors; the penalty for scraping off the slightest excess from any part being the performance of the difficult task of lowering the entire surface to the same extent.

Three surface plates having been thus originated, one of them should be set apart for the production of copies of sufficient accuracy for workshop purposes, and should be allowed to remain entirely undisturbed. When the operator

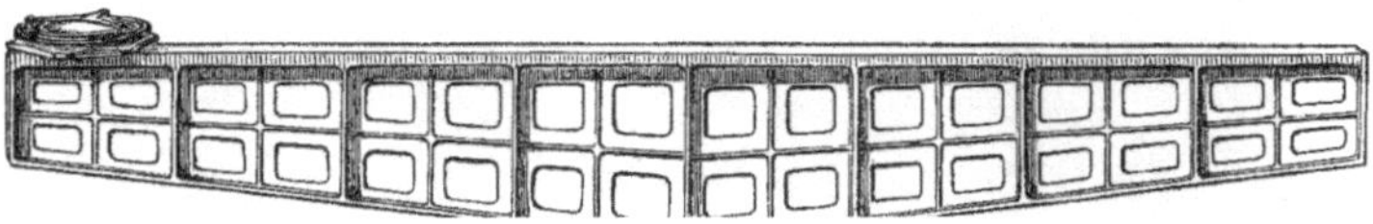

FIG. 14. Cast-iron Straight-edge.

is once in possession of a reliable standard the process of copying is comparatively easy.

The system of scraping and comparison which has given a true plane may be further applied in solving the constructive problem of extending a true plane in any given direction, or, in other words, of making a copy of a plane surface of a larger size than the original from which it is taken. This problem often arises in practice, and an example might occur in the production of a long cast-iron straight-edge, where the surface to be 'trued up' exceeded the standard plate in length but not in width. Such a straight-edge is shown in fig. 14, being a narrow beam trussed at the back in order to ensure the necessary rigidity. Hereafter it will be found that the power of carrying on a plane surface is relied on for the construction of that particular form of measuring machine which we are about to describe.

Those who study geometry cannot but feel some interest in observing the mechanical method of producing (in the sense of carrying onward) a plane or a straight line. Taking the cast-iron straight-edge as an illustration, the surface to be operated on is an elongated rectangular table formed by planing the upper side of the flange of a rigid beam of cast iron. The casting must be supported so that the whole of its surface shall be as nearly as possible in one horizontal plane. The workman commences by filing and levelling the metal to a general average of truth, using ordinary trued-edges, as preparatory to the final testing and correction by a standard plate.

The operation is analogous to that which has been already described. The surface of the true plane is coated with a thin film of reddened oil and is rubbed on the straight-edge at one end only, the object being to obtain an exact copy of the plane on the area covered by it. After a short length has been rendered true, the workman advances the plane, step by step, through intervals which are never greater than one-third of its length, and by repeated comparison and scraping he reduces the added portion to the same continuous level as that from which he started. As he goes on he may find that some portion of the work is below the prolongation of the primary governing plane; if so, nothing remains but to lower the whole at the starting-point and begin again; and indeed some such penalty may naturally be anticipated unless the preliminary testing and correction with a straight-edge has been performed with great care. When the part in advance lies above the true plane, it is apparent that any excess may be removed by this system of comparison; the only thing to be dreaded is the discovery of a hollow portion, which may compel a repetition of the process from the commencement.

We pass on to describe the method of producing a rectangular bar, which may be taken as the type of bars hereafter to be used in the measuring machine, and regarding which it is only necessary at present to say that each pair of opposite sides are truly parallel, and that any two adjacent sides are at right angles to each other.

For the operation now before us we require to work with *three* bars and *two* surface plates. One side of each bar is got up to a true plane, and the respective bars are laid parallel to each other on one of the plates, the corrected faces being in contact therewith. The upper surfaces of the bars are then tested and brought up to truth by laying the second plate upon all the three bars and by scraping the surfaces until an exact coincidence is obtained. The upper sides of the bars are now true planes, but it does not follow that they are parallel to the lower sides. In order to test for and obtain this result, the bars are interchanged and their ends are reversed continually until a perfect coincidence with the upper plate is arrived at under every change of position. It must be remembered that the lower sides are true planes resting on a true plane, and it follows that the coincidence of each bar in all positions, to which we have referred, can only result when the upper surface plate is truly parallel to the lower one and at the same perpendicular distance from it throughout; or, in other words, when the upper and lower faces of each bar are parallel to each other.

The next step will be to make two adjacent sides of a bar at right angles to each other. Here we rely on Euclid's definition of a right angle, and endeavour to make the adjacent angles, formed by the common surface of two bars with a true plane, equal to each other. Take a true plane, A B, fig. 15; place upon it three bars, Y, V, W, whose upper and under sides are parallel true planes. Make *a b* a true plane, and bring *c d*, *e f* to coincide with it by scraping. Turn V over, as in fig. 16; and, remembering that the inverted side is a true plane resting on a true plane, test for a coincidence between *a b* and *c d*. If it be perfect throughout the angles at *a*, *b*, *c*, *d*, are each of them right angles. If not, the error will appear to be twice as great as it really is, and the

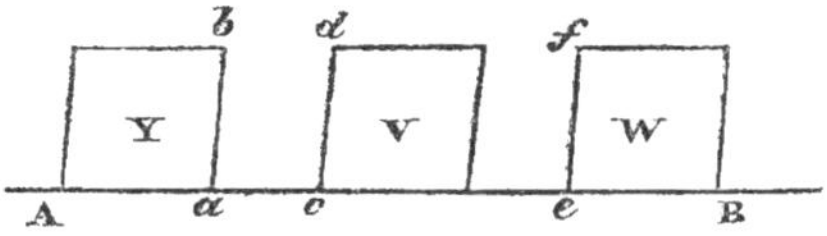

FIG. 15.

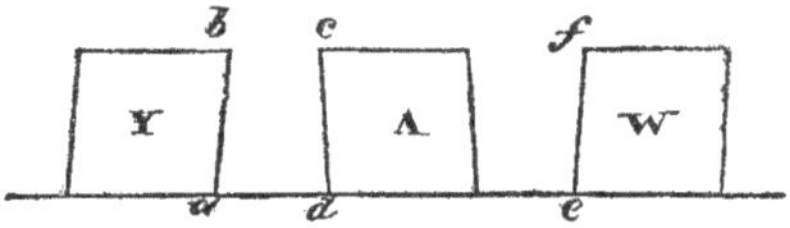

FIG. 16.

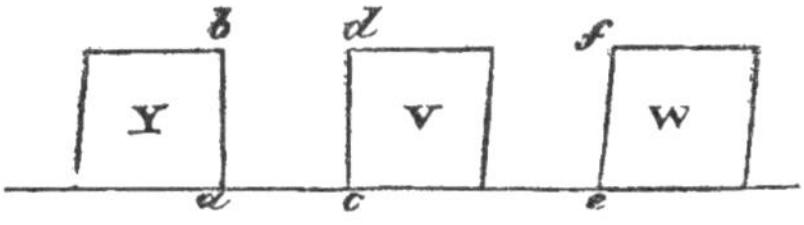

FIG. 17.

operator must remove equal amounts from each side until the contact is found to be perfect in both positions of V, as shown in fig. 17. In like manner the side *f e* of the third bar, W, is tested and made correct by comparison with *a b* or *c d*. The remaining sides—viz. those opposite to *a b*, *c d*, *e f* respectively—may be further subjected to a treatment identical with that already described, after which all the angles of the bar will be right angles.

Or, if preferred, the remaining sides may be rendered parallel to *a b*, *c d*, and *e f*, and the same result obtained, by repeating the operation already set forth for obtaining two parallel planes.

We may here mention that Sir J. Whitworth has proposed an elegant method of recording and preserving two truly parallel plane surfaces on a block, which then becomes a standard for reproducing any number of bars with parallel plane sides, and further provides that each bar, when so formed, shall be of definite and unchangeable dimensions.

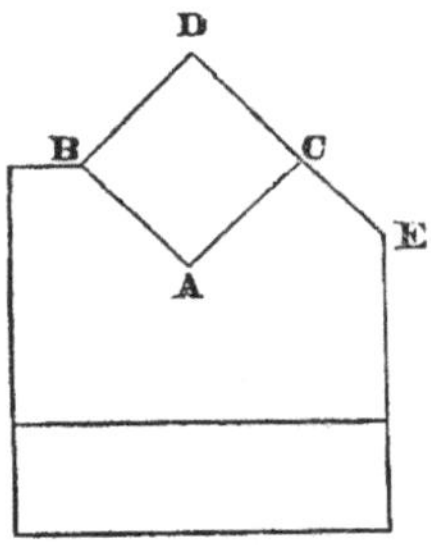

FIG. 18.

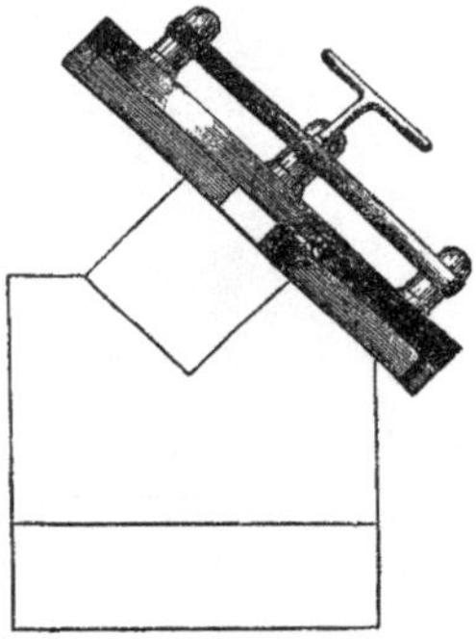

FIG. 19.

A rectangular block of cast-iron, the cross section of which is shown in the diagram, fig. 18, has its upper surface grooved, the faces of the groove in which A B lies being true planes. A rectangular bar, A D, whose opposite sides are true planes, is laid in the groove, and care is taken to obtain a perfect coincidence between one side of the bar (of section A B) and the corresponding face of the groove. The other side of the bar, represented by C D, is parallel to A B, and it can be produced upon the oblique surface represented by C E. Thus the block remains a standard for comparison, having two surfaces—viz. those indicated by the lines A B, C E—which are parallel true planes.

Such a block may be also used to test the truth of the surfaces of a rectangular bar. It is only necessary to copy the second face of a bar, shown by A C, and we have a rectangular groove identical with the rectangular bar, or into which it exactly fits. If the bar be turned step by step through a quarter of a revolution, and an exact coincidence be obtained, in each position, between the bar and the groove, it is an indication that each of the angles A, B, C, and D is of the same magnitude, and therefore a right angle. It is now apparent that if a rectangular bar be laid in the groove, and continually turned through a quarter of a revolution, and there is throughout a coincidence between any one side of the bar and the face A B of the groove, as well as between the opposite side of the bar and a true plane lying on the face C E, fig. 19, the bar in question will be an equal-sided square bar, having its faces A B, B D, D C, C A all equal to each other.

It is also possible to form the rectangular bar and its corresponding groove *at one operation*, and it should be stated that each step in the construction of a measuring bar is based upon geometrical facts. The proposition now to be applied has been already glanced at indirectly, and it is this :—that when the interior angles of a four-sided figure are severally equal to each other, each must be a right angle. Recurring to the block whose cross-section is B A C, fig. 20, the planes through A B, A C are brought as nearly at right angles as they can be made by ordinary workshop appliances. A rectangular bar, whose section is A B C D, is also prepared and made as true as possible by the aid of squares and a surface plate. The bar is then laid in the groove and the faces of the groove are brought, by scraping, into coincidence with those of the bar. The bar is next rotated through a quarter of a revolution, so that the angle marked (1) is brought into the position shown in fig. 21 ; and unless the angle (4) be also a right angle there will be no longer a coincidence between the sides of the groove and the sides of the bar. The amount of error which may exist is revealed by the colouring matter, and one-half, by estimation, must be taken both from the bar and the groove. Thus the operation proceeds :—the workman continually rotates the bar, scrapes away the redundant metal, and compares the results, until a coincidence is obtained in every position, and the task is accomplished.

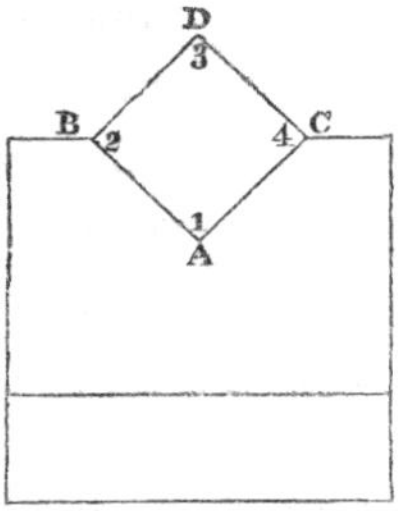

FIG. 20.

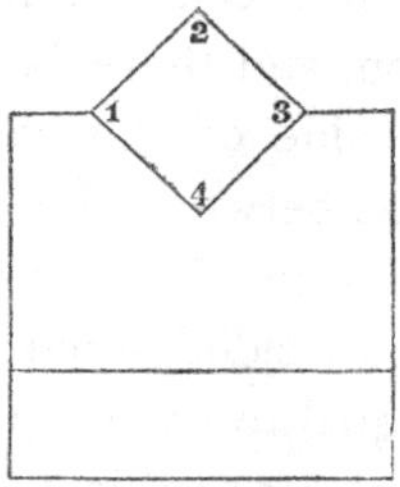

FIG. 21.

Recurring to the mode of extending a true plane in any direction, it is important to remark that when a rectangular bar with sides truly plane is once made, it may be used not only for copying its own form in a groove of equal length with itself, but also for carrying on such a groove indefinitely. And further, if the bar be of sufficient length, it may be made to carry on the rectangular groove with its 'true plane' surfaces across a gap or interval, and to bridge it over without any sensible error. In this way two rectangular plane grooves, separated by a given space, may be brought into as perfect geometrical coincidence as if the separation did not exist, and the sides of the groove were continuous.

Having now obtained a rectangular bar, it yet remains to finish the working up of its ends, and to make the end faces true planes at right angles to its axis. In order to arrive at this result we revert to the rectangular block A B C D E, fig. 22, and make the surface of one end thereof as nearly as possible perpendicular to the line forming the edge of the groove. The bar is laid in the groove and a trial

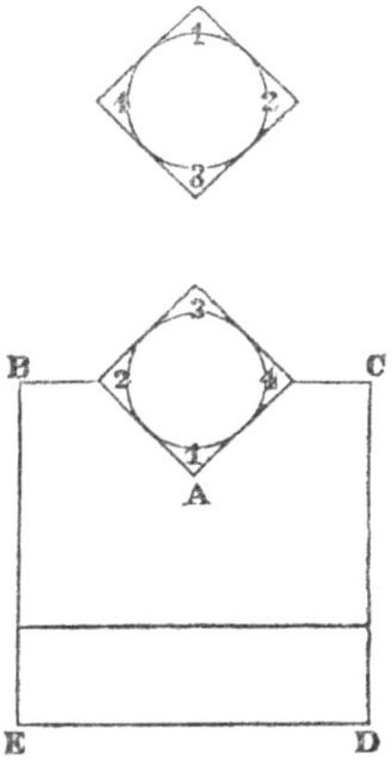

FIG. 22.

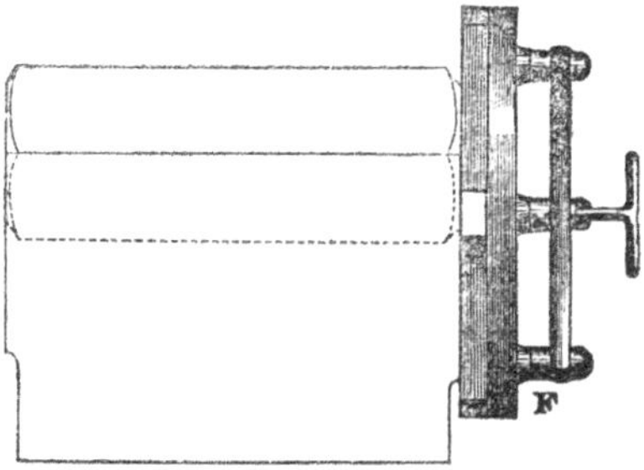

FIG. 23.

plane, F, fig. 23, is brought into contact with the plane end of the block and with the end of the bar. The latter surface is then made a true plane, and occupies the position 1 2 3 4 in the diagram, fig. 22. The bar is now rotated

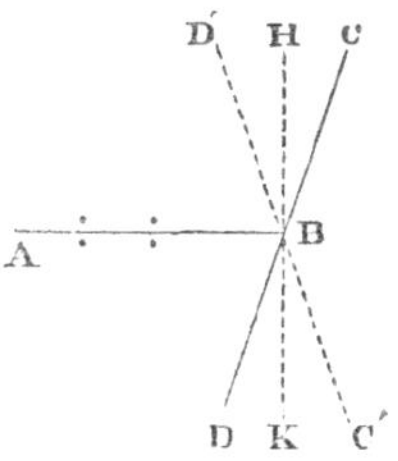

FIG. 24.

through half a revolution, so as to take the position 3 4 1 2, and we have a ready method of testing whether the plane end is or is not perpendicular to the axis of the bar. The geometrical test is at once apparent. If the line C B D, fig. 24, be nearly but not quite at right angles to A B, and if the system A B C D be rotated through 180° about A B as an axis, the lines B C and B D will take the respective positions B C′ and B D′, and the true perpendicular to A B will bisect each of the angles C B D′ or C′ B D. In other words, the true perpendicular lies in the line H K, which is half-way between the two positions occupied by C D. Applying our proposition to the test now under investigation, and remembering that the end surface of the block itself must be worked up to truth during the operation, we shall probably find that one of the angles of the end-face is alone in contact with the trial plane after the rotation. This shows that both the plane A B E D C and the end of the bar are out of truth, and an equal quantity must be taken from each, the directions of the plane surfaces being shifted through half the supposed angle of error. The operation is laborious, as it involves an alteration in direction of the whole surface of a plane. Having brought the trial plane into perfect coincidence with the end of the bar when the angles (1) and (3) are successively uppermost, the same thing has to be repeated with the bar so placed that the angles (2) and (4) are in like positions; and if the coincidence be perfect when the bar is rotated and each angle is successively carried round, the plane end of the bar must be truly perpendicular to its axis.

The final operation is to remove the bar from the groove and rub it lightly on a surface plate.

If it should be thought that the act of rubbing the small end of the bar against the trial surface plate will be competent to hollow out and render untrue the surface with which it comes in contact, we may observe that the area of the surface of the trial plate is, in practice, so much larger than that of the end of the bar, that the abrasion becomes practically imperceptible.

CHAPTER II.

In approaching the discussion of the measuring machine there are two things to be studied, namely, (1) the method of measurement by the microscope and the micrometer, which has been handed down and improved by successive observers from the time of Newton; and (2) the alternative method, known as that of end measurement, which discards the use of the microscope altogether, and relies solely on truth of surface and the sense of touch.

As regards the first method it may be assumed that, so far as our knowledge at present reaches, everything which can be done with line measurement has been fully accomplished. A committee, having among its members Lord Rosse, Sir G. B. Airy, Sir John Herschel, and Mr. Sheepshanks, was occupied from the year 1843 to 1854 in copying or reproducing, according to the system of line measure, bars equal in length to a standard yard. The report of their labours has been published, and shows the extreme care which was bestowed in arranging all the details of construction and observation which were essential to the success of the undertaking. The form adopted for the *standard of length*, and for all its copies, was that of a metal bar, 38 inches long, and 1 inch square in transverse section. A cylindrical hole, $\frac{3}{8}$ inch in diameter, was sunk near each end to a depth of $\frac{1}{2}$ inch; a second small hole was then bored at the bottom of the larger one for the reception of a gold plug, forming a table $\frac{1}{10}$ inch in diameter, on which three fine lines were engraved at intervals of $\frac{1}{100}$ inch in directions transverse to the length of the bar. The measure of the standard was given by the linear interval between the middle transversal line at one end and the middle transversal line at the other end; the part observed in the microscope being at the centre of the line in every case. The other transversal lines were used, in the operations of comparison, for assigning the scales of the micrometers.

The object of engraving the lines which determine the length of a standard on prepared surfaces lying at the bottom of an excavation, instead of on the surface of the bar, is to eliminate the error caused by any accidental small flexure. Thus, the Committee on Standards state in their report that 'if a flexible bar be supported on two points, the extreme length of the bar from the centre of one end to the centre of the other end is not sensibly altered by its flexure, but the distance between two points or lines upon the upper surface may be considerably altered. Bessel has remarked that this objection to line measure is removed if the lines be engraved on surfaces which are depressed to the middle of the thickness of the bar (a principle long since employed in English standard bars); and moreover, the tendency to alter the apparent length, whether at the surface or at the middle of the thickness, may be destroyed by proper adjustment of the points of support, and perhaps still more surely by supporting the bar at numerous points by lever frames, which will ensure equal supporting forces at all the points; or in special experiments by floating the bar in quicksilver.'

The lever frames above mentioned are constructed so as to ensure an equality of pressure at each point on which the bar rests. The principle of the contrivance will be understood from fig. 25. A lever, D C R, with equal arms, and pivoted on a centre at C, has a pair of levers, A B, P Q, each having equal arms, pivoted on centres at D and R.

Since all the arms of the levers on either side of every respective fulcrum are equal, it follows that the pressures supporting the bar at the points A, B, P, Q must necessarily be equal in order to balance throughout the system.

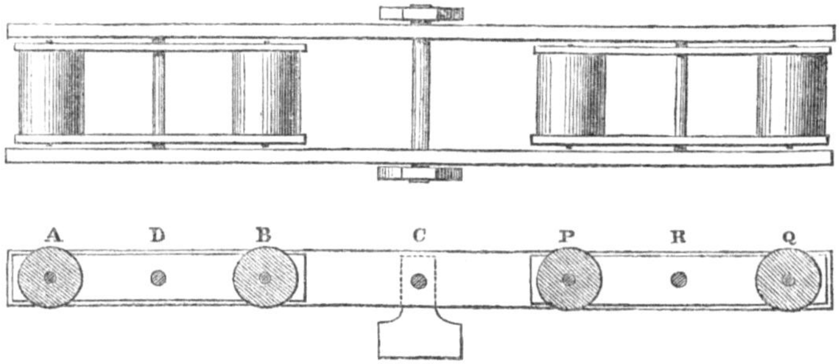

FIG. 25. Lever Frames.

In the extremely accurate observations made by the committee the bars were covered with goldbeater's skin and floated in a trough of mercury, which was surrounded on all sides by a water casing. They were so adjusted as to be under no sensible strain, and remained as nearly as possible free, although prevented from shifting their position in any direction.

The lines traced upon the gold plugs were illuminated by artificial light, and their position was read by micrometer microscopes of the ordinary construction, but supported on solid piers of masonry. It will be evident that the accuracy of line measurement depends altogether upon the observations made with the microscope; and in order to present a clear idea of this branch of the subject we refer to a sketch of the field of view of a reading microscope, No. 4, as used by officers of the Standards Commission for observing the graduations of a standard whose divisions are seen in the middle of the field. The drawing, fig. 26, is taken from page 116 of Appendix VII. to the Fifth Report of the

Commissioners appointed to enquire into the Condition of the Exchequer (now Board of Trade) Standards, A.D. 1871.

It would appear that a movement through one division of the micrometer-head of the microscope used by Mr. Sheepshanks corresponded very nearly to a linear interval of ·000036 inch in the object measured. And the value of a division of one of the microscopes at present in use at the Standards Office is estimated at ·0000318 inch. We may, therefore, say roughly that the microscopes used for

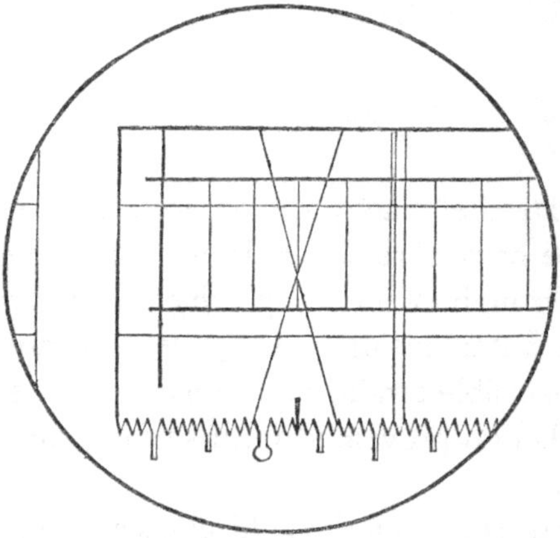

FIG. 26.

comparison of lengths will read to an interval of about $\frac{1}{30000}$th of an inch in linear measure; and further, that this graduated interval may be subdivided according to estimation in the usual manner.

In line measure the errors traceable to fluctuations of temperature in the bars themselves give rise to many complications. Thus, it was stated by Mr. Sheepshanks that an expansion for $\frac{1}{10}°$ Fahr. in a 36-inch bronze bar amounted to one division of the micrometer, or ·000036 inch, and that the expansion for a rise of temperature of $\frac{17}{100}°$ Fahr. produced the same elongation in a like bar of iron. In order to take cognizance of the exact temperature at which the observation was conducted two thermometers were attached to each bar, and in some few instances holes were sunk in the bars themselves, and bent thermometers, with small bulbs, were laid therein. With these thermometers observations were made for estimating to $\frac{1}{100}$th of a degree Fahr. the specific temperature at which copies of the standard yard represented its length correctly.

It is no part of our purpose to magnify unduly the difficulties which were encountered in the observations of length by line measure, or to throw any doubt upon the 200,000 micrometer readings which were made during the progress of the work. Our only object is to trace exactly what was done, and to point out to what extent the actual observations made are to be relied on in a trial of strength between different systems. We may, however, point out that in line measurement the observer endeavours to recognise in the field of view of the microscope the exact coincidence of two cross-threads with an engraved line on the gold plug, or he may note the coincidence of two parallel lines. He thinks he sees the coincidence, and he records it; but, in truth, the judgment of one observer seldom agrees with that of another, and the exact reading sought for eludes the grasp just at the instant when it might be thought to have been fairly secured.

Anyone who has read the graduations of a divided limb with a microscope, when taking an astronomical observation, will acknowledge and appreciate the difficulty. The engraved line is magnified till it becomes a coarse furrow—the cross threads or wires are enlarged so as to appear of sensible thickness, and to lose all resemblance to a spider's web; and the observer soon becomes conscious that he is powerless in the attempt to subdue the errors traceable to the rough outlines which are presented by the most delicate objects when unduly magnified.

In truth the nature of the difficulty inseparable from an attempt to define the exact bisection of the image of an engraved line on a bar by the image presented by the cross wires in a microscope is very clearly stated both by the Commissioners for Standards in their report to Parliament, and by Mr. Airy in his paper on the work of the Commissioners as read before the Royal Society.

Thus we find (Report of the Commissioners for Constructing New Standards of Length, page 8) that 'Mr. Sheepshanks proceeded to make a series of comparisons of a particular bar (bronze 28) with numerous bars, in order to select five bars which in respect of the distinctness of their engraving, of their floating evenly in quicksilver, &c., might seem well adapted to be taken as Parliamentary standard copies. When the observations had been carried to an extent which, it was supposed, would be perfectly satisfactory, it appeared that the result of comparisons by different observers was sensibly different. An extensive series of new observations by numerous observers was at once commenced. The discussion of these observations showed clearly, as had been suspected before, that there is a difference among the results of different observers far exceeding the uncertainty inferred by the theory of probable errors from the observations of each observer taken separately. In other words, each observer has a *personal equation* in the microscopic observation of the engraved lines. The existence of discordances of similar character between the astronomical observations of different observers has long since been recognized. The origin of this personal peculiarity is very obscure, but the fact seems to be beyond doubt.' Then came the question of the amount of the probable error of a single observation of an observer, and here we may refer to Mr. Airy's paper (Phil. Trans. vol. 147, page 671). 'In the autumn of 1852, the feeling of insecurity with regard to personal equation, and the influence of the bar's position, pressed so strongly on Mr. Sheepshanks, that he determined on making an extensive series of observations with the assistance of observers who were familiar with micrometrical observation, and whose accuracy could not be questioned.' Their names were Henderson, Dunkin, Simms, W. Simms, jun., De la Rue. They were set to the task of comparing each of the five bars above referred to with 'bronze 28,' and the results are stated in a table in terms of an interval in length corresponding to the reading of the micrometer head, where one division (or one-hundredth part of a revolution) represents ·0000358 inch. Mr. Airy goes on to state that the probable error of a single observation by Mr. Sheepshanks is 0·55 of a division, that is ·00001969 of an inch, or about $\frac{1}{50000}$th of an inch.

The observers are next compared with each other, and we are told that the probable errors of the different observers as compared with those of Mr. Sheepshanks are :

Henderson : Sheepshanks as 9 : 8
De la Rue : Sheepshanks as 6 : 7
W. Simms, jun. : Sheepshanks as 5 : 8

and so on, the conclusion being that 'no sensible error would be committed by assuming the observations of all the observers to be equally good, except those of W. Simms, jun., one of whose observations is equivalent to two of any other observer.'

Mr. Airy then gives an abstract of the result of the comparisons of the five bars with bronze 28, and tabulates the discrepancies. Taking the comparison of bronze 10 with bronze 28, it is stated that, as the mean of 142 observations, Mr. Sheepshanks estimated bronze 10 as being shorter by 1·28 divisions; Mr. Dunkin, with 30 observations, made it shorter by 1·41 divisions; Mr. W. Simms, jun., with 24 observations, found it shorter by 1·83 divisions.

The discrepancy between Mr. W. Simms and Mr. Sheepshanks amounted therefore to ·55 of a division, or to ·55 × ·0000358 inch, or to ·00001979 inch, and the remark made is, 'It seems difficult to resist the conclusion that there is a very sensible personal equation.'

The report first quoted adopts these numbers and sums up as follows:—'In the comparison of bronze 10 with bronze 28, there is a discordance between the results obtained by Mr. Sheepshanks from 142 sets of observations and by Mr. W. Simms, jun., from 24 sets of observations, amounting (in ten-millionth parts of an inch) to 198.'

Now $\frac{198}{10000000}$ inch is very nearly $\frac{200}{10000000}$ inch, or $\frac{1}{50000}$ inch, and this statement is therefore an admission that after 142 observations by Mr. Sheepshanks, and 24 observations by Mr. W. Simms, jun., there is a discordance in the comparison of the lengths of the bars, bronze 10 and bronze 28 amounting to the coarse interval of $\frac{1}{50000}$th of an inch—a magnitude which a skilled workman supplied with a bench measuring machine would detect in a moment when required to apply end measurement to a standard bar. For it should be remembered that in Sir Joseph Whitworth's workshops the men are habituated to work to what is termed *one division*, or $\frac{1}{10000}$th of an inch, and that it requires no particular effort to subdivide one graduation of the divided wheel of a measuring machine into five intervals recognisable to the eye; and further, that a difference of one-fifty-thousandth of an inch thus ascertained is appreciable to the sense of touch by observing the difference of tightness when a cylindrical gauge is passed between the bars of the measuring machine. The manner in which this is done will be fully explained hereafter.

We have next to examine the method which the Commissioners adopted in obtaining standard end-measure bars. In their Report, page 8, while referring to Bessel's preference for an end-measure bar where the whole length of the bar is adopted as the measure of length, and discussing his reasons for this preference, they observe :—'A second reason assigned by Bessel is, that the principle of end-measure is more convenient than that of line-measure for the production (that is, practically, the comparison) of copies of the standard. It appears probable that this remark is well founded as regards the construction of secondary standards for *commercial purposes*; but it is doubtful whether it applies to secondary standards for *scientific purposes*; and it can scarcely apply to *primary* standards, in regard to which the consideration of convenience (in the few comparisons that may be made at intervals of many years) has no weight, when contrasted with the consideration of the conservation of the standard length. Another reason (assigned also by Bessel) is, that it is more convenient for use; but the members of the Committee, who have witnessed the operation of measuring a geodetic base by means of bars carrying the line-measure, have been led to form a high estimate of the convenience of the line principle in that instance. It is, moreover, to be observed that the whole of the British geodetic bases have been measured by bars constructed on the line principle; and a standard which is intended to apply advantageously to them must be constructed on the same principle. The end-measure has never, so far as we know, been applied to any scientific determination in England.' And again, in page 13, the Committee state :—'From the commencement of our labours we have proposed to ourselves the formation of a standard *à bouts* equivalent in length to the Parliamentary standard *à traits*, as necessary for the completeness of our undertaking. After consideration of the methods of verifying and using such a standard, we recommend that the metal be steel, and that the ends be of hardened steel, or hard stone, as quartz or sapphire; and that the end surfaces be curved, forming portions of one sphere whose centre is in the centre of the bar.'

In Mr. Airy's paper, page 685, there is a detailed account of the formation of end-measure bars with spherical ends. It is there stated :—'The form given to the end bars depends in some degree upon the process adopted for comparing end-measure bars with line-measure bars.' The influence of the form given to the end of a bar upon the method of measuring its exact length is then explained, and it being obvious that the microscopic reading can only be applied to the observation of engraved lines, we can readily understand that some special contrivance must have been arranged. The idea carried out was to measure the interval between the centres of two bars placed in one line, and then, after turning each bar through 180°, to take the measurement over again. In this way, by operating with three bars taken in pairs, the true length of one end bar was deduced from a series of double measurements of pairs of bars. The process has been the following :—

Conceive that two end bars are prepared with plane ends, and that a small well-hole or excavation is made at the middle of each bar, so as to lay bare the axial line, and let a gold plug be inserted at the bottom of the hole. Upon the plane surface of the plug, which is axial to the bar, a transverse line is engraved, just as in the case of an ordinary line-measure bar.

The bars A B, B C are now placed as in fig. 27, end to end, with their axes in the same straight line, (1) with the shaded ends away from each other, and (2) with those ends in contact. Micrometer microscopes are employed to read the interval between the engraved lines, and we have for the first reading,

$$b + c = \text{length read by microscope,}$$
$$= l \text{ suppose,}$$

and for the second reading,

$$a + d = \text{length read by microscope,}$$
$$= l' \text{ suppose.}$$
$$\therefore\ a + b + c + d = l + l',$$

or A B + B C $= l + l' =$ a known quantity, whence the sum of the length of a single pair of bars is determined.

Conceive further that the same operation is extended to

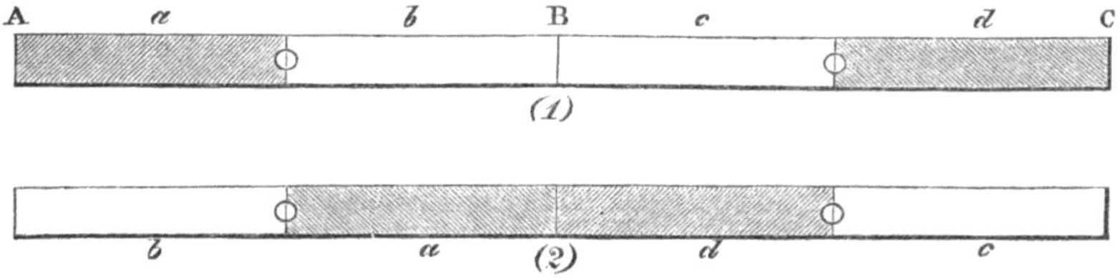

FIG. 27.

a third bar, and that it is repeated upon each pair of bars formed by placing in contact two out of the three.

Let x, y, and z, be the respective lengths of the bars; by the above method of repeated observation we obtain

$$x + y = \text{a known quantity,}$$
$$x + z = \text{the same,}$$
$$y + z = \text{the same,}$$

three simple equations for determining x, y, and z. It follows, therefore, that by an indirect method we can ascertain the precise length of an end-measure bar without placing its ends under the microscopes at all; but the operation would fail if the bars were not placed with their axes in exact coincidence with the same definite line. As to the possibility of adjusting the bars with their axes in one line, and as to the form of the end surface, Mr. Airy proceeds as follows :—'In the opinion of the Committee, the use of *plane ends* could not be sanctioned. One difficulty attending them is, the uncertainty as to the removal of particles of dust, &c., from every part, and the certainty that if any such particles remain they will disturb the measure of the bar; while at the same time the act of pressing two bars together, either with or without a slight movement, does not tend to remove the particles. Another difficulty is that, even though we were assured that the ends are perfectly clean, yet we cannot be assured that the bars, when laid in a horizontal position end to end, will touch in the whole breadth of their ends; inasmuch as, if the position of one of the bars is slightly erroneous in azimuth, or slightly erroneous in inclination to the horizon, no force which it is practicable to apply in the mounting of the bars would tend to correct the error. The only way which seems practicable is, to place the bars in the vertical position, one bar standing freely upon the end of the other; but this requires a rather inconvenient arrangement of the microscopes.

'These errors and inconveniences are entirely avoided by giving to the ends the form of a spherical surface, whose centre is the centre of the division-line at the middle of the bar's length. There is a tendency to thrust away particles of dust from the point of contact; and a small error in the elevation, or inclination, or azimuthal position, of either bar produces no error on the distance between the centres of the division-lines of two bars.'

That is to say, end-measure bars must not have flat ends; for there is no known method of placing them with their axes in the same straight line, except perhaps that of allowing one to rest endwise on the top of the other. And the best form is a bar with spherical ends, for in that case the measurement of the length of a bar can be effected without placing the bars accurately in such a position that their axes coincide in direction.

The latter proposition becomes very apparent when explained more fully, and will be understood from fig. 28. Conceive that two bars with spherical ends are placed in contact at P, with their axes somewhat inclined; and let E, F, be the centres of the bars. It is evident that E P F is the sum of the two radii E P, and P F, and that $EF = b + c = l$.

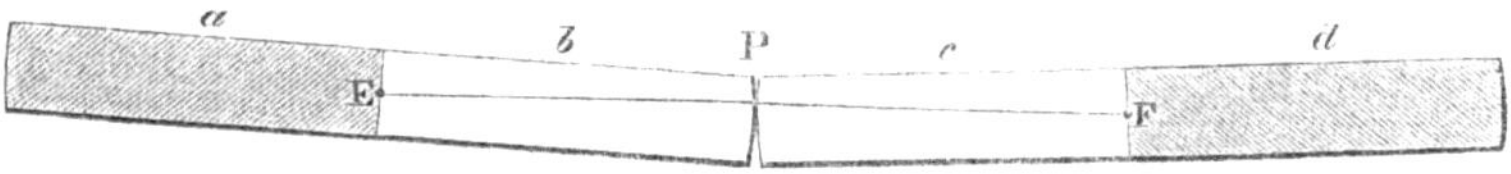

FIG. 28.

Next, place the bars with the opposite ends in contact, as shown in fig. 29, when we again have $EF = a + d = l'$. In other words, we repeat the measurements with inclined bars just as effectively as in the former instance where the bars had plane ends. It is true that the transverse lines at E and F which are viewed in the microscopes are somewhat

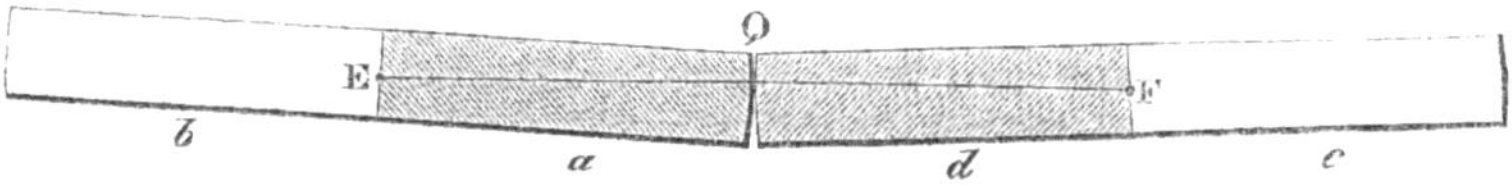

FIG. 29.

inclined to each other, but the central portions of these lines are bisected by the cross-wires of the micrometer, and the reading is practically the same as if the engraved lines were strictly parallel.

As regards measurement with plane ended bars, it may be convenient to preface our enquiry by taking a simple illustration which exhibits the ordinary practice of the workshop. We may instance the case of a workman provided with a pair of callipers, fig. 30, and required to apply himself to the task of gauging and comparing two cylinders, for the purpose of making one an exact copy of the other. Calling the original A, and the copy B, the workman would adjust the callipers until he had obtained a certain degree of tightness when passing them over the surface of A—he would then take every care to preserve the measurement unaltered. It should be understood that the ends of the callipers are blunt edges, and that in bringing the copy B to the size of the original A the workman gradually reduces B, until, to his sense of touch, the feeling of the tightness of the instrument when passed

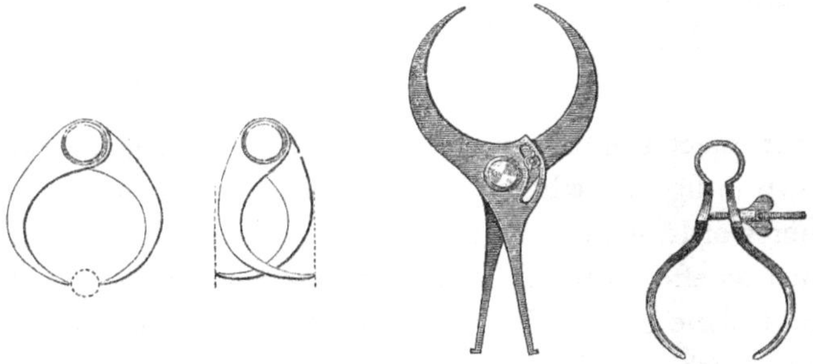

FIG. 30. Various Forms of Callipers.

over the one is the same as that experienced when gauging the other.

The comparison of size is a delicate operation, and one in which the conclusion formed as to the identity of two things may be more or less erroneous, by reason of the difficulty of holding the legs of the callipers in a plane exactly perpendicular to the axis of the cylinder tested. It is perfectly apparent that any twisting or inclining of the plane in which the legs of the instrument lie must affect and disturb the conclusion of the observer. The transverse section of a cylinder is a circle, whose diameter gives the measure of the cylinder, but an oblique section of a cylinder is an ellipse, whose major axis is greater than the diameter of the cylinder. When the callipers are held so as to embrace the transverse section the observation is correct; when they embrace the opposite arcs of an oblique section the observation is not that sought for, and is worthless. Although a skilful workman may use callipers so as to obtain very good results, it is evident that he struggles against an inherent defect which cannot be remedied.

In like manner callipers have been constantly employed for the conversion of line into end measure. In effecting this interchange the workman lays his callipers upon the graduations of a foot-rule, and reads off the interval as nearly as he can; he then shapes some piece of metal which is to be made of given dimensions until the callipers pass over it with an estimated degree of tightness, and he thus transfers an interval on the rule to some part of a solid body. In doing so, the sources of error are threefold, and one is very serious. It is impossible to determine how nearly the end of each leg of the callipers coincides with the central line of the somewhat coarse graduation of the rule; and if that were better known there is still the particular difficulty which has just been commented on, viz. that of holding the instrument in a plane truly perpendicular to the surfaces under measurement; and one other, which has not yet been mentioned, arising from the uncertain spring or yielding of the legs of the instrument under the variable pressure of contact.

Some few years ago, and before the construction of special appliances, the common method adopted in the workshops for producing an object of any definite size was that given above, and the accuracy of the work done depended mainly on the personal skill of the workman. There was another difficulty experienced when it became necessary to compare work done at different places, for there was no such thing known as a standard inch, and it was by no means certain that the graduated foot-rules, from which the measurements were derived, were exact copies of one and the same standard, or that the divisions upon them were correctly given.

Thus the whole subject of measurement remained in a state of uncertainty and confusion until Sir Joseph Whitworth applied himself to the construction of an end measuring machine. Before commencing that task he had satisfied himself that the only practicable mode of measurement suited for the workshop should be one founded on truth of surface and the sense of touch—the delicacy of the nerves of feeling being, in fact, a thing quite disregarded and neglected by all others who had applied themselves to improving mechanical measurement; and it has been in developing to the fullest extent the results obtained by the contact of surfaces which were almost ideal in their excellence that a system has grown up which stands quite alone and removed from all thoughts of rivalry or competition. Now that the result has been made known, and we are able to assign the value of minute differences in size which were previously only guessed at, as something hopelessly beyond the reach of exact observation, it is perhaps difficult to realise the view which the subject presented in the first instance, and before the problem had been solved. But, nevertheless, the true way of learning a new thing is to attempt this realisation, and the enquirer may with advantage put himself in the position of one beginning from the standing-point of what was known before the improvement was originated, and then working step by step to its final accomplishment.

And, indeed, those who thus approach the subject for the first time can scarcely fail to be interested in a notice of experiments which demonstrate the possibility of making certain minute differences of size which are simply felt, and are not seen, a measurable quantity. We are so much in the habit of defining all minute linear intervals by the aid of some form of microscope that it is a new thing to seek for the power of measurement anywhere but in optical contrivances. It is the originality of the conception that we can measure without distinctly bringing into the field of vision the thing measured which gives the system now under discussion an interest that it would not otherwise possess.

We have spoken of the method of gauging by means of callipers, with their blunt and somewhat uncertainly shaped points; and the use made of the sense of touch has been referred to as giving the workman the power of estimating the size of a cylinder by his perception of what he felt in passing the callipers over it. Conceive now that the callipers are replaced by two parallel true planes, fixed at a distance equal to the diameter of the cylinder, and, further, that these planes can be caused to approach or separate by almost insensible intervals. We should probably be first of all struck by observing that many gradations of increasing tightness can be felt and recognised as differing from each other while passing the cylinder between the planes. First the contact may be just felt, then some slight resistance is experienced, which gets more and more evident, until a certain degree of pressure must be exerted in order to pass the cylinder between the surfaces. The degree of tightness is a matter of estimation, and no two observers would be in exact agreement as to its amount. At the present time, however, we can measure the interval corresponding to any difference in this feeling of tightness, and are enabled thereby to form a new and reliable idea of the meaning of a mechanical fit between two pieces of metal. For example, let the planes be so adjusted that it is just possible to feel the contact as the cylinder passes between them, and then approach the planes by $\frac{1}{40000}$th of an inch. A sensible degree of tightness will become apparent, and the increase of pressure necessary to overcome the resistance due to the closer grasp of the planes against the cylinder can be distinctly appreciated. But the object in view is exact measurement, and the question before us is, how to supply something which shall always indicate one exact degree of tightness, and no other, in a mechanical fit between the particular surfaces in contact.

So far as the case we have taken—viz. that of a cylinder passed between two true planes—is concerned, we must now abandon it, as we require the contact of continuous surfaces in order to present the illustration in a simple form. It will lead to more certain results if we conceive a standard inch bar, with parallel plane ends, to be passed between the true planes forming the terminations of the legs of the imaginary pair of callipers. But the observation must be conducted in a different manner, and according to the altered condition of things. It would not be possible to move the bar between the planes, as the slightest twisting might cause it to become locked, and the method adopted is, therefore, to leave the bar *in situ*, that is, with its end touching one of the true planes; and further, to employ an additional instrument, called a feeling-piece, for testing the nature of the contact.

The annexed drawing, fig. 31, shows a feeling-piece (drawn to full size), being a plate of steel, with parallel plane sides, and having slender arms, one for its partial support and the other for resting on the finger of the observer. The state of things is now as follows:—One true plane is fixed, and one end of the inch bar rests in contact therewith; the other true plane is movable, and its exact position is determined by a mechanical multiplier, which is capable of advancing it by intervals of, say, $\frac{1}{1000000}$th of an inch. The feeling-piece, sometimes called a gravity-piece, is placed between the end of the inch bar and the movable true plane, and provision is made that each arm of the feeling-piece shall rest on its table. The operator now lifts one arm, with

FIG. 31. Elevation and Plan of Feeling-piece.

his finger, fig. 32, allowing the other arm, C D, to rest on the table, whereby the weight of the feeling-piece is partially supported. The movable plane is in such a position that a certain degree of tightness is felt between the surfaces when the arm is thus lifted; and the question will be—how

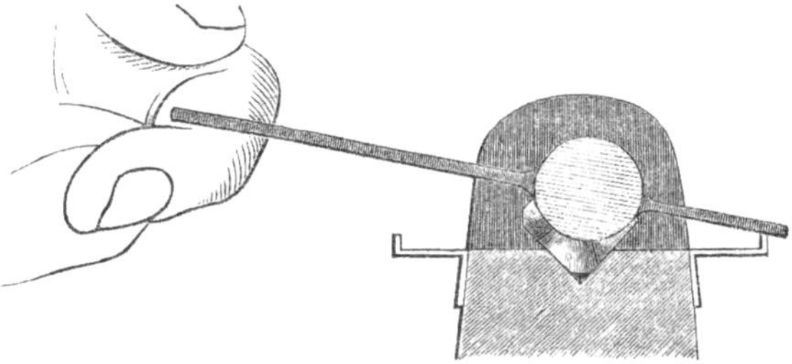

FIG. 32. Method of manipulating with Feeling-piece.

much? It is agreed only to seek for that degree of resistance which shall be just sufficient to support the feeling-piece, and to prevent it from sinking down again when the finger is removed. In order to attain the object in view Sir Joseph Whitworth has constructed a so-called mechanical multiplier of marvellous delicacy, which gives the smallest increase of step by step movement in the advancing plane that can be safely relied upon. The millionth machine derives its name from the multiplier, which is constructed so as to diminish or enlarge the space between the measuring planes by $\frac{1}{1000000}$th of an inch for each graduation on the rim of a divided circle, while the less delicate or workshop machine will, in like manner, give movements differing by $\frac{1}{10000}$th of an inch.

Having thus given a general idea of the method of making an observation upon this new principle, we reserve an account of the complete instrument with its millionth multiplier for the next chapter.

CHAPTER III.

It will be readily understood that the measuring machine, as constructed, is rather a machine for comparison than for simple measurement. It is not employed as a measurer in the ordinary sense of the word; that is to say, we could not take a specimen bar to the machine, *per se*, and placing it therein, without other aids, ascertain the precise length of the bar. The apparatus is to be regarded as a machine for measuring very minute differences in the lengths of bars specially prepared for end measurement. It is therefore peculiarly adapted for multiplying copies of standards of length.

The Millionth Measuring Machine for the comparison of inch-bars is shown in perspective in fig. 33, and in detail in fig. 34 on a scale one-third the full size; and again in fig. 35 on a scale one-half the full size. It consists of a rigid cast-iron bed, A, having two heads, resembling lathe-heads, which are prepared for the reception of the sliding bars B and C. The upper part of each head forms a cover to its sliding bar, and is secured to the bed of the machine by screws. The sliding bars are rectangular, with faces made truly plane, according to the process described in Chapter II. They terminate at the measuring ends in circular planes, about $\frac{3}{10}$ inch in diameter, and the utmost care is taken to ensure that these plane ends shall be truly perpendicular to the respective axes of the bars. This is the turning-point of the whole operation; and those who have questioned the advantages of end measurement have disbelieved in the possibility of this construction—they have said that it is impossible to make a bar with the ends perpendicular to its axis. The answer is that a trial plane may be shown to be in perfect contact with a fixed plane, and with the plane ends of a rectangular bar in each of the four positions assumed when the bar is rotated through 1, 2, 3, 4 right angles from its normal position in a rectangular groove. The matter has been already discussed at page 24.

The bed of the machine has a right-angled V groove between the heads, corresponding with the sliding bars, the faces of the grooves being true planes at right angles to each other; whence it follows that the sliding bars have a bearing throughout the whole extent of the surfaces in contact; and further, that the axes of the bars, when slid along the groove, will move either in coincident or parallel lines, and will thus fulfil the essential and indeed imperative condition that the plane ends shall be truly parallel throughout the motion. The function of the grooves becomes, therefore, as important as that of the bars, and the excellence of the machine depends on its rigid fulfilment of the geometrical conditions now set forth.

We have next to describe the construction of the mechanical multiplier for advancing the end of one sliding bar through spaces differing by ·000001 inch (one millionth of an inch). Taking the rectangular sliding bar C, we find, by referring to the sectional drawing in fig. 42, that it is bored out from end to end, and that the measuring end of the bar is formed at the extremity of a cylindrical plug, having a shoulder, up to which it is screwed. Also that the sliding bar is caused to traverse in the V groove by the rotation of a square-threaded screw, having twenty threads to the inch, and passing along the centre of the bar.

The screw in question is connected with the bar by a

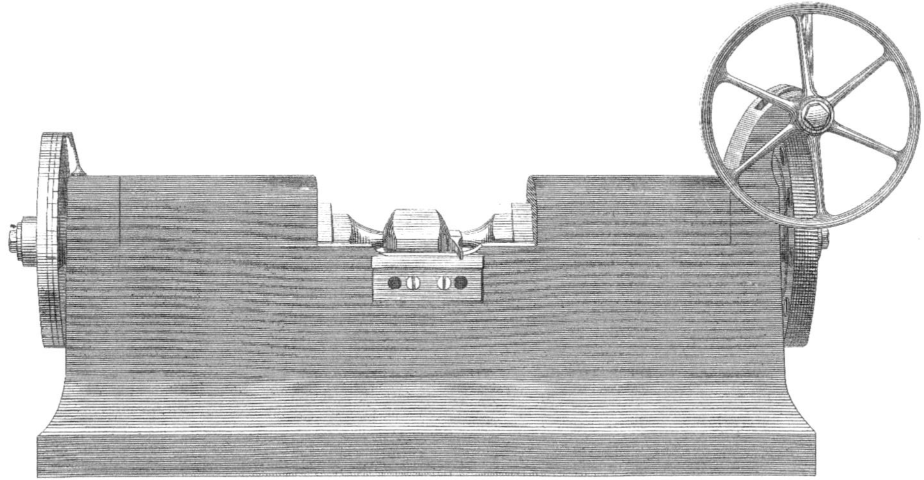

H Adlard sc.

FIG. 33. – PERSPECTIVE VIEW OF WHITWORTH'S MILLIONTH MEASURING MACHINE.

London: Longmans & Co.

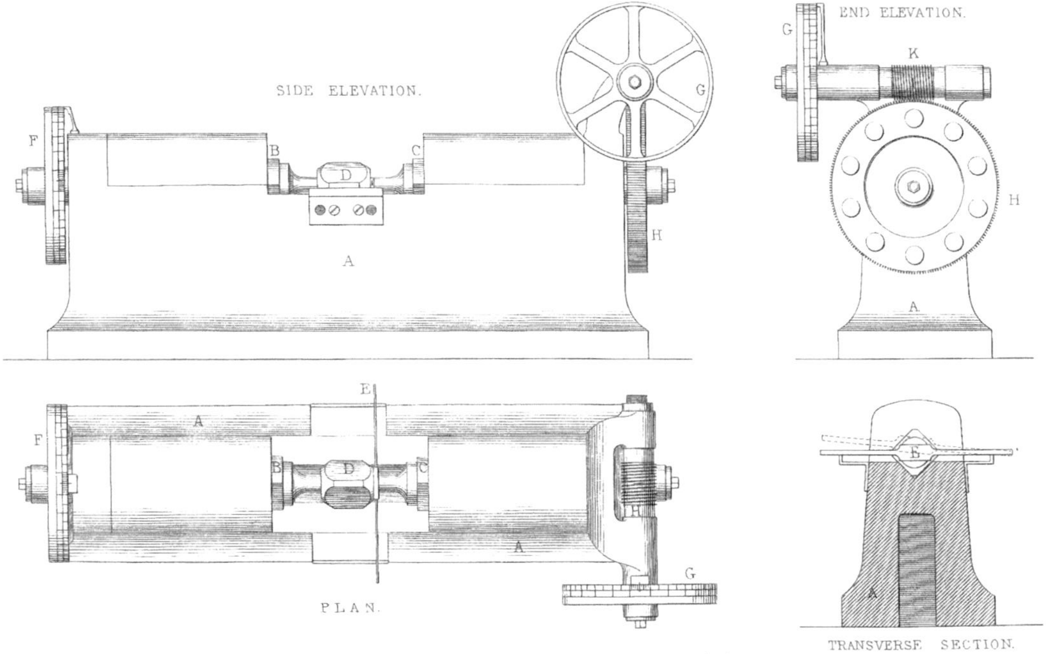

FIG. 34. GENERAL VIEWS OF WHITWORTHS MILLIONTH MEASURING MACHINE. ($\frac{1}{3}$)

H. Adlard sc.

London, Longmans & Co.

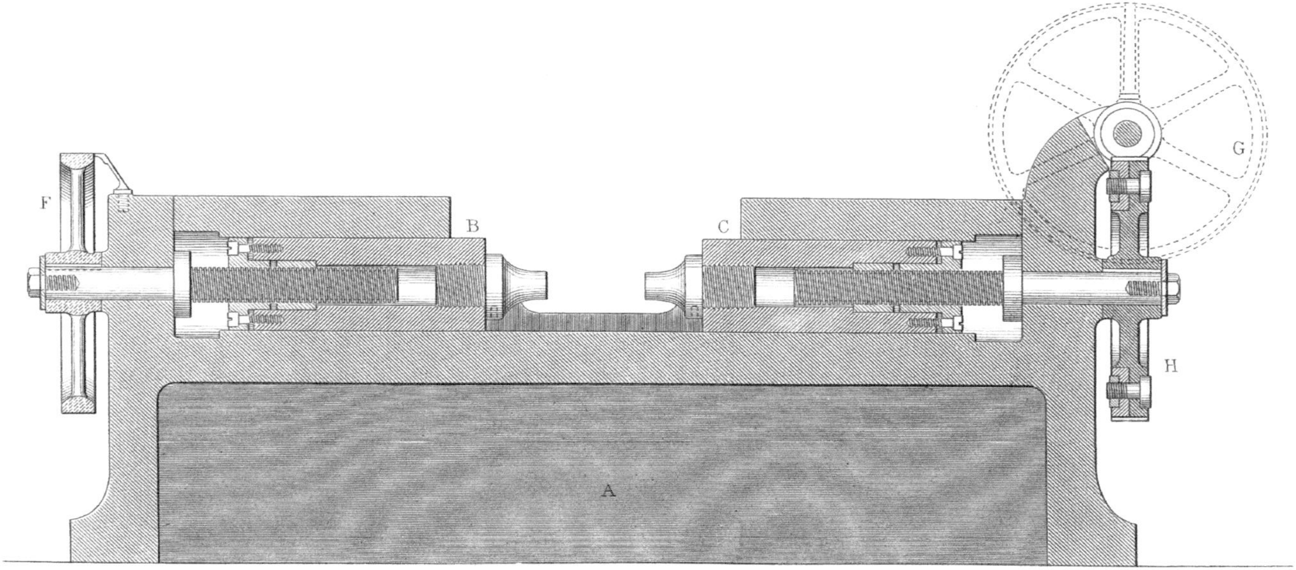

H. Adlard sc.

FIG. 35. LONGITUDINAL SECTION OF WHITWORTH'S MILLIONTH MEASURING MACHINE. ($\frac{1}{2}$)

London: Longmans & Co.

divided nut, which is sunk into a recess at the end of the bar. The construction of the divided nut is important, and is the following:—After a nut has been fitted as truly as possible upon the screw it is cut into two lengths, one of which is provided with a flange. The two portions are threaded upon the screw, a small space being left between them; they are let together slightly, and are made to grip the threads of the square-threaded screw in both directions by the four screws which pass through the flanged portion of the nut, and are tapped into the end of the rectangular sliding bar, the base of the other portion of the nut bearing closely upon the bottom of the recess in the bar. Such a contrivance prevents the slightest play or backlash in the screw, and provides the means for tightening it up at any time, to compensate for wear.

The slight pressure brought to bear upon the measuring planes when the machine is in use is taken by the solid collar on the screw, the shoulder of which bears against the inside of the bed of the machine.

The screw is rotated by a screw-wheel, H, having 200 teeth, which again is caused to rotate by a tangent screw, K, carrying a wheel, G, with a micrometer graduation of 250 divisions upon its circumference.

Now, $$20 \times 200 \times 250 = 1{,}000{,}000.$$

Whence the rotation of the micrometer head, G, through one division will advance the measuring bar, C, through one-millionth of an inch.

The tangent screw is also made in two halves, upon the same principle as the divided nut, and is adjusted to press both sides of the teeth of the screw-wheel so that the backlash or loss of time is reduced to a minimum. Sir J. Whitworth states that he has sometimes found the backlash to lie within two millionths of an inch.

The mechanical multiplier above described requires to be supplemented by a coarser apparatus, and accordingly the second sliding bar B is also advanced by a screw with twenty threads to the inch, the connection between the screw and the bar being similar to that already described; but the wheel F, fixed on its screw, has 250 divisions, and there is no further multiplication on the side of B. A rotation of F through one division therefore causes the advance of B through $\frac{1}{5000}$th of an inch, and the final adjustment is given by the bar C. It ought to be well understood that the object of a microscope is simply to increase the extent of surface which the eye traverses in measuring a distance, and thereby to render minute differences of length appreciable; but the extent to which it has been found that this can be carried by means of microscopes has been demonstrated to be limited and uncertain as compared with the method of end measurement. In the measuring machine a motion of ·000001 inch (one millionth of an inch) in the screw is represented by nearly ·04 inch (four hundreths of an inch) upon the circumference of the dividing wheel, so that the machine magnifies 40,000 times, and the eye traverses over 40,000 times the space that is being measured. These approximate numbers very clearly exhibit the function and purpose of a mechanical multiplier.

The gravity-piece E rests at each end upon the projecting edges of the bed, and is moved vertically by the finger between the ends of the standard bar D and of the sliding bar C as shown in the plan and transverse section (fig. 34); it is lifted at one end into the position indicated by the dotted lines in the section, and at first falls freely when released; the bar C is then gradually advanced by means of its screw until the pressure of contact is sufficient to support the weight of the gravity-piece, which presently remains suspended between the ends of the bars when lifted, and has lost the power of falling by its own weight, although still capable of moving freely when touched. The exact position of the wheel G is noted by the graduations on its circumference; and the standard bar D is removed, the bar C being drawn back by the wheel G. If it be required to produce a duplicate of the standard, the proposed duplicate is put into the machine, its ends being made true planes at right angles to its axis; one end of the duplicate is placed in contact with the end of the bar B, which has remained unmoved since the last measurement; the gravity-piece E is lifted up by the finger and allowed to fall between the other end of the duplicate and the end of the bar C. The bar C is gradually advanced by means of the wheel G, until the fall of the gravity-piece is arrested; the reading of the wheel then indicates whether the proposed duplicate is of exactly the same length as the standard, or shows the difference of measurement to millionths of an inch.

In tracing the successive steps in the progress of the construction of end measuring bars, we have to point out that the first millionth measuring machine was adapted for the reception of a bar 36 inches in length, and that a standard end-measure bar was obtained from a comparison with line-measure copies of the Government standards in the manner and by the apparatus described subsequently at page 56.

Such a machine, as applied to a standard yard, was shown by Sir Joseph Whitworth at the Exhibition of 1851, and obtained a Council medal. A further notice of the instrument is to be found in the Report of the Select Committee of the House of Lords appointed to consider the Weights and Measures Bill, where it is stated that Sir Joseph Whitworth, on June 7, 1855, gave the following evidence: 'I may mention that, in order to avoid any effect which might arise from a bias of my mind, I have frequently employed another person to set up a bar by a millionth of an inch at a time unknown to me, and looking myself to the gravity-piece, I have arrived at the same result. This experiment I made with my standard yard bar in the presence of Lord Rosse and Mr. Babbage, at the Great Exhibition in 1851. It does happen that after many repeated trials the parts get a little less. This piece of steel, which is soft, we find wears, and if a number of experiments are made the micrometer has to be moved forward a little more to compensate for it. The millionth of an inch is so small a quantity that you have only to rub a piece of soft steel a very few times to diminish its thickness a millionth of an inch.'

A standard yard of end-measure having been obtained from line-measure, it became a comparatively easy task to subdivide it into feet, and for this purpose three bars were prepared, each a little longer than one foot. A temporary abutment was then raised in the bed of the measuring machine between the measuring planes, the distance between the measuring plane on the sliding bar and the fixed measuring plane forming the abutment being slightly in excess of the length of one of the bars, together with the thickness of the feeling-piece. The shortest of the three bars was then placed in the groove of the machine, with the feeling-piece in position. Before disturbing the sliding bar the reading on the micrometer-wheel was carefully noted. The movable measuring plane was next brought into contact with the feeling-piece until the latter was just supported, and the micrometer-wheel was again read before bringing it back into its original position. The bar was then removed and replaced by the second one, the movable measuring plane being again brought into contact with the feeling-piece, until it was just supported. The micrometer-wheel was again read, and the difference between this reading and that noted when the former bar was in the machine gave the excess of length of the second bar. The excess of length was carefully reduced by the methods already described, until by repeated trials the second bar was brought to the exact length of the first one, the test being that the readings on the micrometer-wheel were identical for the two bars. The same course was followed with the third bar, whereby the three bars were brought to one and the same length; but it did not follow that each bar was exactly one foot in length. In order to test for such equality the micrometer-wheel was turned back to the graduation observed when the standard yard was in the machine, the temporary abutment was removed, and the three bars were placed in the groove of the apparatus with their ends abutting against each other; their aggregate length was tested with the gravity-piece, and, the behaviour of the gravity-piece turning out to be the same as when the standard yard was in the machine, it followed that each one of the bars was exactly $\frac{1}{3}$ of a yard, or one foot long. By the same means a foot was subdivided into inches, and a standard bar one inch long was obtained.

The standard inch bar (for comparison) is shown in fig. 36; it is nearly a counterpart of a measuring bar, and is

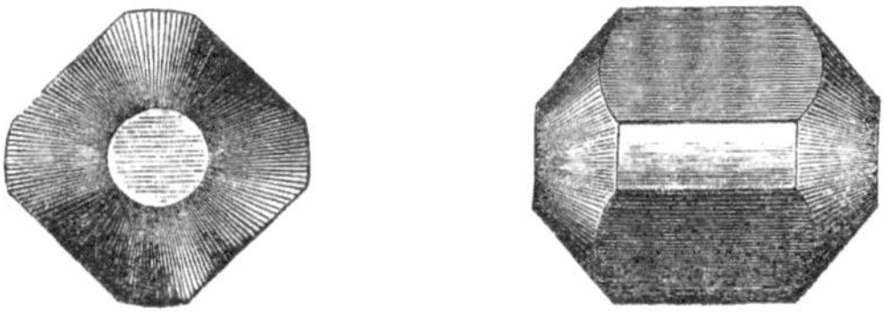

FIG. 36. Standard Inch Bar for End Measure.

drawn to full size. It lies in the same rectangular groove as the bars, and its plane ends are truly parallel to the measuring planes.

A smaller machine—being that now presented to the

reader—has also been constructed for comparing and copying inch bars, and the capabilities of the machine as a difference measurer were demonstrated by Sir J. Whitworth at a meeting of the members of the Institution of Mechanical Engineers at Birmingham, on July 27, 1859.

At the close of the meeting Sir Joseph showed that an advance of ·000001 inch ($\frac{1}{1000000}$th of an inch) was distinctly indicated by the gravity-piece becoming suspended instead of falling; and the turning back of the divided wheel through two divisions, representing ·000002 inch, was then sufficient to cause the gravity-piece to drop, and included consequently all the play in the four bearings of the two screws and two collars. Sir Joseph Whitworth showed also that the fineness of measurement obtained by the machine was sufficient to detect the expansion in length of an inch bar caused by a momentary touch of the finger, the bar then measuring ·000001 inch longer than previously (the expansion of iron being about $\frac{1}{150000}$th of its length for each degree Fahr., a rise of temperature of $\frac{1}{7}$th of a degree expands an inch bar $\frac{1}{1000000}$th of an inch). He stated that in his larger machine for measuring the standard yard, with a bar 36 inches long, the same amount of expansion was shown by the momentary contact of the finger-nail.

The finest measurement required the precaution of freedom from dust and moisture in the atmosphere, and from any current of air interfering with uniformity of temperature; and the machine was therefore kept in its glass case during the time of use, with an opening only sufficient for moving the micrometer wheel and lifting the gravity-piece. By sufficient care in these respects the measure of a space corresponding to half a division on the wheel, or $\frac{1}{2000000}$th of an inch, has been rendered distinctly perceptible.

We have now to refer to the influence of temperature upon the subject of measurement. For example, suppose that a square bar of steel, 36 inches long, and with plane ends at right angles to its axis, is prepared as a standard of length, and that exact copies of this bar are required. It is beyond question that Sir Joseph Whitworth has demonstrated the possibility of producing copies of such a standard which shall not be affected by direct errors of observation in a degree at all commensurate with those which vitiate line measure. If it be asked what is the measure of the difference between the two, the answer would be, that repeated experiments have shown that the machine employed for comparison will measure intervals of $\frac{1}{1000000}$th of an inch, and that the extreme limit of possible accuracy is estimated by Sir Joseph Whitworth at one millionth of an inch. Leaving any general considerations for a future period, we pass on to consider the mode in which errors due to alterations of temperature are got rid of.

It should here be borne in mind that changes of temperature have affected the results of line measurement to such an extent that we may almost say that the Committee have practically failed to produce mechanical copies of their standard. At page 8 of their Report they state: 'It is found that the length of one yard, as given by the lost Imperial standard, is represented with no sensible uncertainty by the following bars, with the temperatures placed opposite to them:—

Bronze 19, or No. 1, at 62·00° Fahrenheit.
Bronze 20, or No. 2, at 61·94° ,,
Bronze 2, or No. 3, at 62·10° ,,

And so on.

That is to say, bronze 20 will be of the same length as bronze 19, if by any means you can, having brought the first bar to an exact temperature of 62° Fahr., reduce the temperature of the second bar by $\frac{6}{100}$ of a degree below 62° Fahr. Unless this be done the two copies will not be identical in length. It may be that such a mode of comparison is sufficient for scientific purposes, but it cannot, without a mere perversion of language, be said that bronze 20 is a mechanical copy of bronze 19. If it were so, the two bars should be always of the same lengths at identical temperatures.

But now let us turn to the system of end measurement. There is no natural standard of length adopted in this country, and therefore the selection of a standard is a matter of arbitrary choice. Granted that a certain bar of metal is a yard long at a given temperature; place the bar at the disposal of a mechanician, and he can provide a mechanical copy which shall always be identical with the standard bar at identical temperatures. All that is necessary is that the bars and the machine should be kept as nearly as possible at one and the same temperature. A bar can be copied at one temperature as well as at another, and the only danger is that the temperature of the machine should differ from that of the bars, so as to cause them to contract or elongate when placed *in situ*.

If a copy be identical with the standard at any temperature at which the observation is made, and the standard is one yard long at 62° Fahr., the copy will be also one yard long at 62° Fahr., and will satisfy the requirements for which it has been produced.

With regard to the selection of 62° Fahr. as the normal temperature, Mr. Airy said, in his evidence before the Select Committee on the Weights and Measures Bill (see p. 17 of the Report, dated 1855): 'There is no particular value in 62°; any other temperature might do, but there is a particular reason for having something not extremely different from it, and it is this: the most delicate operation which we have to perform is the measurement for the bases of surveys, and that is done in the open air, in summer generally, and the temperature is not far distant from 62° Fahr.'

It further appears from the same Report that Sir Joseph Whitworth stated (p. 5): 'With regard to the temperature of the standard, which is fixed at 62°, I am of opinion that it is too low, and that it should be nearer that of the human body, on account of the difficulty experienced whilst the work is in operation, through its being affected by the heat of the operator. No inconvenience would arise to the workman by a temperature of from 70 to 80°.' And again, at p. 15, he said: 'There is no reason why these standards, and the room in which they are kept and in which the copies are verified, should not be maintained always at one uniform temperature, night and day. Mr. Appold keeps his house so by a thermometer, which regulates a valve through which the gas flows, and the temperature is uniform to within 1°. I have no doubt that a room might be always kept at one temperature, and in that room the standards ought to be kept, and, of course, the copies verified. I think it is a great mistake to have the standard temperature at 62°.'

As a further illustration of his meaning, Sir Joseph Whitworth said: 'I may mention that a bar like this standard size, at the temperature of 62°, I could show the Committee that if I touch it for an instant with my finger I have lengthened it. The heat is our only difficulty. We have none in making the ends perfectly parallel, nor in producing a perfect straight line; that is to say, we have it correct to the one-millionth of an inch. The only difficulty we have to contend with is the heat, and that I would provide against by making the standard temperature higher.'

Notwithstanding the efforts of the Commissioners to restore and reproduce a standard of length, there can be no question that, so far as relates to the standards used by engineers, and which determine the lengths of an inch and a foot in our workshops, the practical standard has been originated by Sir Joseph Whitworth. The system of gauges, both external and internal, which he introduced has been universally accepted; and if at the present time an engineer wanted a spindle one inch in diameter he would assuredly be quite content to copy a Whitworth inch standard gauge.

The origin of these gauges is the following:—In 1834 Sir J. Whitworth set himself to obtain an end measure bar, one yard in length, which should as nearly as possible coincide with the Government standard. The line measure bars which he possessed for comparison were standard yards, one made by Donkin, and the other by Troughton and Sims. Each bar was of brass, and the yard was measured between two engraved lines. The instrument used by Sir Joseph Whitworth for copying these standards is represented in fig. 37, where microscopes are placed over each end of the line measure bar. The microscopes are fixed in their position by clamping screws, and one bar is capable of being replaced by a second one, the transverse lines on the latter being compared with the original, and the difference being measured and registered by the observer.

After the observation has been completed, and the cross wires of the micrometer have been adjusted in the correct position, the line-measure bar is removed and an end-measure bar is substituted in its place. The operation of reduction of length of the end-measure bar has to be gone through until the wires of the microscope coincide with the edge of that vertical plane end of the bar which takes the place of the engraved line.

Those who are acquainted with the practice of the workshop know perfectly well that a workman is so habituated to rely upon the sense of touch in estimating a mechanical fit that it is as hopeless as it would be undesirable to provide

FIG. 37. Whitworth's Microscope Apparatus for the Comparison of Bars.

him with a microscope and to ask him to estimate a difference of size between two objects with such an instrument. To do so would be to require him to replace a single observation by a double one. If the microscope is to be applied, there must be an observation of both ends of the piece under measurement. Unless the object is exceedingly minute it cannot be comprised within the field of view, whence it follows that the instrument and its adjuncts must be carried bodily from its one position of measurement to another before a result can be arrived at. With a measuring machine one observation only is required, and Sir Joseph Whitworth has stated that he could discern when blindfolded a difference of size corresponding to a movement of $\frac{1}{50000}$th of an inch in the measuring planes of the ordinary workshop machine. As a matter of fact it is easier to work to the ten-thousandth of an inch from standards of end measure than to one-hundredth of an inch from the engraved lines on a graduated rule. In all cases of mechanical fitting the measures of length employed should be prepared gauges or bars, and not ordinary divided rules.

CHAPTER IV.

THE reader who has followed us through the description of the millionth measuring machine will readily comprehend the construction and object of the Workshop Measuring Machine, which has been designed for enabling manufacturers to refer at any time to accurate fixed measures, and thereby to give them the power of reproducing identical copies of pieces of machinery, as well to preserve a system of sizes of the various fitting parts unaltered.

The practical value of the workshop measuring machine consists in the facility which it affords for the construction of difference gauges.

For example:—Every external cylindrical surface having to work in an internal surface should have a certain difference of size; and the amount of that difference can only be determined by close observation, and by reference to the results of experience. Take the case of a railway axle. If the bearing in which it works be too small, the friction set up by rapid rotation will develop heat; and if, on the other hand, the bearing be too large, the wear will be excessive, or we may even say that the effect is the same as if the part were worn out before starting. It is, therefore, most important that in all instances where rapid rotation and great strains have to be undergone, the proper difference of size, when it has been once ascertained by experience, should be adhered to with strictness.

Two gauges will in this way prove to be of great service while carrying out the manufacture of an axle, the journal being made to a standard gauge, and the bearing being bored out so as to fit a difference gauge somewhat larger in size, and of the precise difference in diameter which experience has shown to be necessary, regard being had to the conditions under which the axle is to work. It may therefore be conceded that every manufacturer should be in a position to manufacture his own difference gauges.

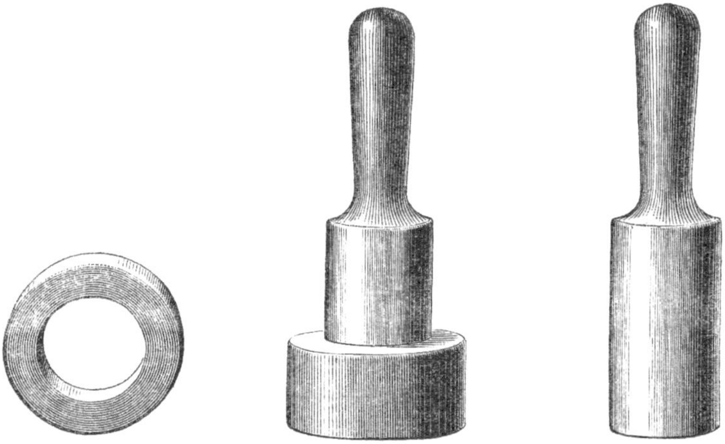

Internal Gauge. External Gauge.

FIG. 38.

The gauges to be used for measuring both internal and external diameters are made in pairs, being called respectively *external* and *internal gauges*; they are shown in fig. 38. The standard cylindrical gauges are usually made from $\frac{1}{10}$th to 2 inches diameter, but they are also made of larger diameters; they are a necessary adjunct to the workshop measuring machine for making difference gauges. Sir Joseph Whitworth employs cylindrical external and internal difference gauges in the manufacture of rifle-barrels; each gauge differs by $\frac{1}{5000}$th of an inch in diameter, and a workman is enabled to feel his way step by step, and thus to make the bore of any number of barrels or tubes identical throughout.

These gauges are finished with great care, and are made true after being case-hardened. They are so hard that scarcely anything will abrade them, except the grinding process to which they have been subjected.

If we revert to the practice of fifty years ago for an additional illustration, it may suffice to state that before the period of exact measurement thousands of spindles in a cotton factory were each separately fitted into the bolster in which it respectively worked. At the present time all such spindles are made to gauge, and are interchangeable. Or, again, we may refer to the manufacture of small arms, where it is no uncommon thing to put together the parts of a rifle in less than three minutes, although the separate pieces have never been brought together before, but have been taken by chance from the lots made of each kind.

It cannot be impressed too forcibly, both on the student of mechanics and on the workman, that accuracy of measurement is essential for good and efficient workmanship, and that it tends to economy in all branches of manufacture to make the separate parts interchangeable.

With a view to providing an apparatus which shall possess the practical utility that has now been pointed out, the machine represented in a perspective view by fig. 39 has been constructed. The mechanical multiplier is of much lower power than in the case of the millionth machine, and consists of a simple graduated micrometer wheel, actuating a screw of 20 threads to the inch. The value of one division on the micrometer head is no longer $\frac{1}{1000000}$th inch, but is reduced to $\frac{1}{10000}$th inch. Not only is there a diminution in the magnifying power of the instrument, but the ends of the sliding bars are separated by a gap or interval which may be enlarged so as to receive circular

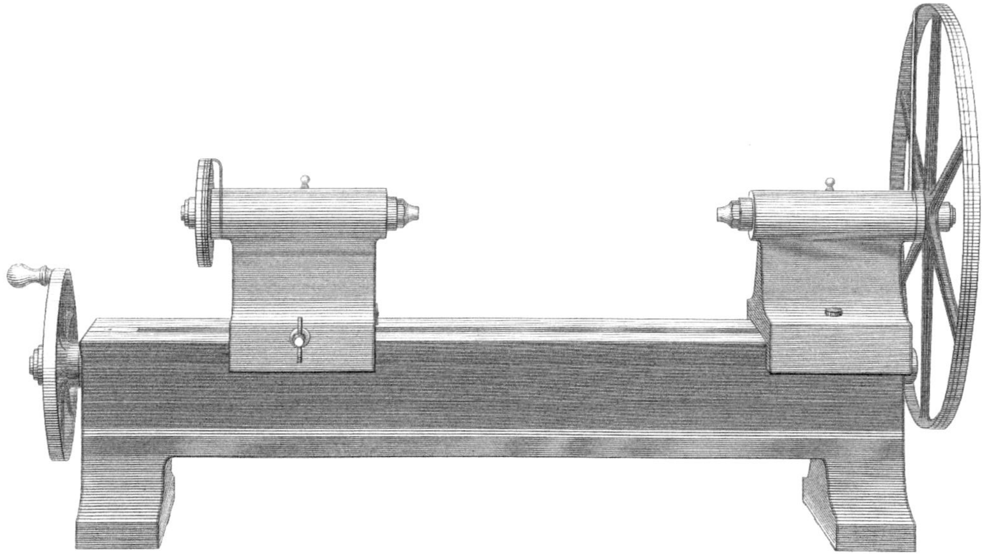

H. Adlard sc.

FIG. 39. – PERSPECTIVE VIEW OF WHITWORTH'S WORKSHOP MEASURING MACHINE.

London: Longmans & Co.

gauges up to 6 inches in diameter, and other pieces up to 12 inches in length. Unlike the millionth machine, it is not adapted for the comparison of prepared bars by the aid of a feeling-piece, but the observations are simply estimations of size by reference to the degree of tightness felt in passing the object between the plane ends of the sliding bars, and no feeling-piece or other adjunct is employed for determining the exact amount of the resistance thus experienced. It will, therefore, be seen that the capabilities of the workshop machine are far more extensive than those of the other and more sensitive instrument; and although its performance may be less calculated to excite surprise and admiration, it is better adapted to the fulfilment of every-day requirements.

A plan and front elevation of the machine now under consideration is shown in fig. 40 to a scale one-fifth the full size. The general appearance of the apparatus is that of a small turning-lathe; there is a cast-iron bed A formed of two parallel cheeks, connected at the ends, and stiffened by intermediate ribs; the bed itself being supported upon standards or feet placed at the respective ends. The whole is cast in a single piece; and in order to guard against the want of steadiness and the risk of distortion of the bed arising from the unevenness of the workman's bench, the under surface of the standard at A′ has two narrow longitudinal fillets, forming feet or supports, while a single fillet attached to the opposite standard forms another foot on that side, and thus the machine rests on three feet, just as if it were an ordinary surface plate. The upper part of each cheek is spread out into a flange, with one inclined and one vertical face, so as to form together a guide for the headstocks B, C, which are carried upon the bed, and contain the adjustable planes D, E, between which the objects to be measured are passed. The vertical and inclined surfaces of each flange are carefully scraped and made true planes, and the same operation is performed on the corresponding surfaces of the dovetailed grooves at the foot end of each headstock.

One headstock C is immovably secured to the end of the bed by the screws *a a*; but the other headstock can be slid along the bed by means of a handwheel F attached to a quick-threaded screw G, which lies between the two cheeks of the bed and is supported on cylindrical bearings at each end. There are nuts *b*, *b*, screwed so as to bring the shoulder of the collar on the screw to bear against the face of a boss at the opposite end of the bed, and thus to provide against the occurrence of any endlong motion in the screw.

The screw G is connected with the movable headstock B by means of a gun-metal block, which is recessed into its under side, and secured by tapped screws passing upwards through the flanges of the block into the foot of the headstock. The block has an internal thread, through which the screw passes, so that by turning the screw the headstock is either advanced or drawn back, and any backlash which there may be in the movement (due to the difference between the thickness of the thread on the screw and the width of the female thread in the block) is taken up by a flanged nut which is recessed into the block, the device being the same as that already described for traversing the sliding bars in the millionth measuring machine. The square-threaded screw G is cut with a double thread of $\frac{1}{2}$ inch pitch, the hand-wheel attached to it being $5\frac{3}{10}$ inch in diameter.

The adjustable planes D, E, carried by the headstocks are advanced or drawn back by an arrangement similar to that adopted in the sliding headstock of a lathe; and, since the details are the same in both headstocks, it will suffice to describe the traversing head B.

A hole is bored from end to end of the upper part of

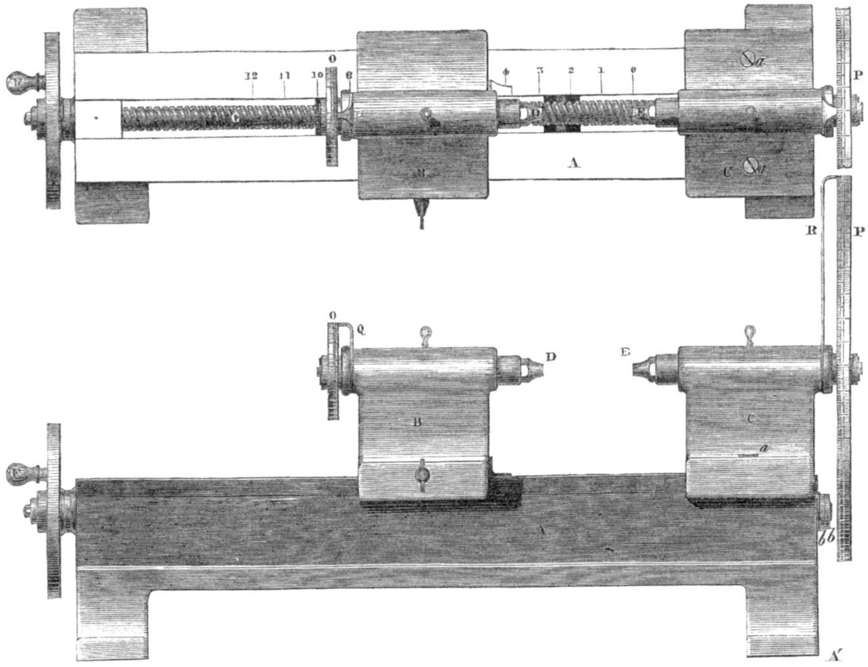

FIG. 40. Plan and Elevation of the Workshop Measuring Machine ($\frac{1}{5}$).

the headstock, great care being taken to make it truly cylindrical, and with its axis parallel to the bed of the machine. It is also essential that the cylindrical bore in the one headstock should exactly coincide in position with the corresponding cylindrical bore in the opposite headstock, or, in other words, that the two internal cylindrical surfaces should have a common axis.

A cylindrical sliding-piece is then fitted into the bore, the diameter of the sliding-piece being slightly less than that of the internal surface, according to a difference previously determined, and gauged by difference gauges. The cylindrical sliding-piece is also itself bored throughout, and one end thereof is closed by a cylindrical plug, which is screwed in by means of a six-sided head. Beyond the head there is a circular true plane, forming one of the two measuring surfaces on which the value of the instrument depends, the object in view being to determine the interval between two circular true planes, whereof one is carried by each headstock.

Each cylindrical sliding-piece is prevented from turning round in the headstock by a narrow key which forms a feather, and is recessed, partly into a longitudinal groove cut along the under side of the sliding-piece, and partly into the headstock itself; the length of the groove determining the distance to which the measuring plane projects beyond the headstock. The feather is retained in its position by a covering plate secured to the headstock by a tapped screw.

The cylinder, with its measuring plane, is caused either to advance or recede by a square-threaded screw with 20 threads to the inch. The screw passes through a split nut fitted into a slotted hole in the cylinder, and the two halves of the nut clip the screw by four screws tapped into the lower half of the nut. A solid collar is turned upon one end of the screw, one side of the shoulder bears against a flanged stop, which is fixed to the headstock by four tapped screws. A graduated hand-wheel O is keyed to the outer extremity of the screw spindle, and secured by a plate and tapped screw. A wheel P, corresponding to O, but of larger diameter, is attached to the screw at the side of the fast headstock, C. Index pointers Q, R, are attached to each of the headstocks by the screws which pass through the flanges of the stops.

The graduated portions of the machine are three in number. In the first place, intervals of 1, 2, 3 ..... 12 inches are marked upon the surface of the bed, so that the movable headstock can at once be approximately placed in any required position; secondly, the small wheel, 3 inches in diameter, carried by the movable headstock, has 250 divisions round its circumference, each complete revolution of the wheel corresponding to a backward or forward movement of $\frac{1}{20}$th of an inch on the part of its measuring plane, and the movement through one division on the rim advancing or withdrawing the measuring plane $\frac{1}{5000}$th of an inch; and, lastly, the large wheel of the fixed headstock, which is 11·8 inches diameter, has 500 graduations upon its rim, and therefore gives ·072 inch nearly for each division, so that the mechanical multiplier in this case is 741 nearly.

In the comparison of rectangular bars a simple form of support between the headstocks is essential; but in the case of cylindrical gauges, or other portions of work, the operation can be easily effected without such assistance, the gauge or article being merely held in the hand and passed between the ends of the measuring planes, as shown in fig. 41. Large gauges and pieces which are too heavy to be conveniently held in the hand may be suspended vertically above the machine: the adjustment can then be given correctly, as in the case of the very smallest articles. When the true adjustment has been given to the machine it has been found that an observation of $\frac{1}{40000}$th of an inch in the distance between the measuring planes causes a distinctly perceptible increase or decrease in the resistance which the cylinder or other object encounters in passing between them.

In order to explain clearly the method of using this machine, we will suppose that an external cylindrical difference gauge, 4·003 inches in diameter, is under construc-

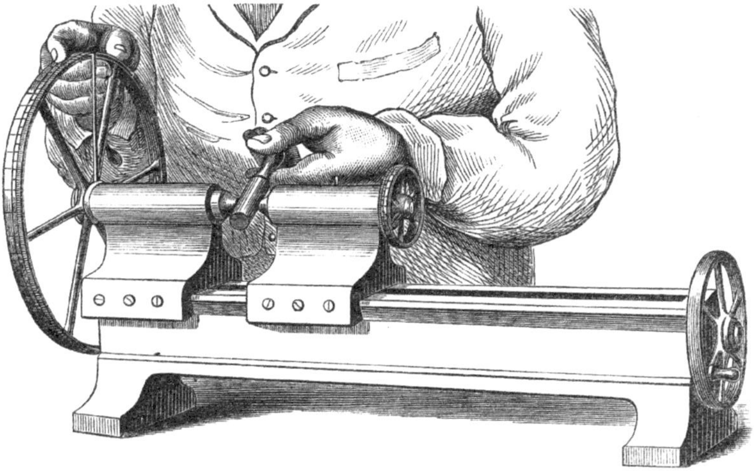

FIG. 41. Method of manipulating with the Workshop Measuring Machine.

tion; or that a workman is required to produce a gauge which shall exceed the standard 4-inch gauge by $\frac{3}{1000}$ths of an inch. The movable headstock of the machine having been clamped at the 4-inch division on the bed, its wheel is adjusted till the 4-inch standard gauge can just pass freely between the ends of the sliding bars, the largest wheel upon the fixed headstock having been previously set and read off. This wheel is then moved through 30 divisions, which gives the exact difference required in the distance between the sliding bars. Perfect but free contact with their ends on the part of the proposed 'difference gauge' will then prove its correctness.

During the progress of the competitive trials of guns before the Special Committee at Shoeburyness, in 1864, Sir Joseph Whitworth designed a contact measuring instrument for the purpose of ascertaining the enlargement, if any, which might take place from time to time in the bore of his 70-pounder gun.

The instrument, when completed, afforded a ready means of testing the strain upon guns according to a principle commonly carried out in the testing of girders; that is to say, the observation was directed to the determination of the amount of permanent set produced, and gave immediate warning of the occurrence of any undue strain. It would appear that many rifled guns have been permanently injured by being subjected to proof with excessive charges of powder, and the object kept in view by Sir Joseph Whitworth has been to ascertain by measurement the amount of charge which any particular description of gun will bear without a permanent alteration of its parts. It may here suffice to state that one division of the scale on the steel tube represents one ten-thousandth of an inch in enlargement of the bore.

The observation is conducted as follows:—Before commencing the proof an exact measure is taken of the bore of the gun (see fig. 42), which shows a plan of a 7-inch gun, the dotted lines indicating the measuring instrument A in its place in the bore; the whole being drawn to a scale one-twentieth of the full size. A separate drawing shows the projectile which, according to results obtained by Sir Joseph Whitworth, has the most perfect form for flight. The proof is begun by firing a powder charge somewhat smaller than that employed during the ordinary

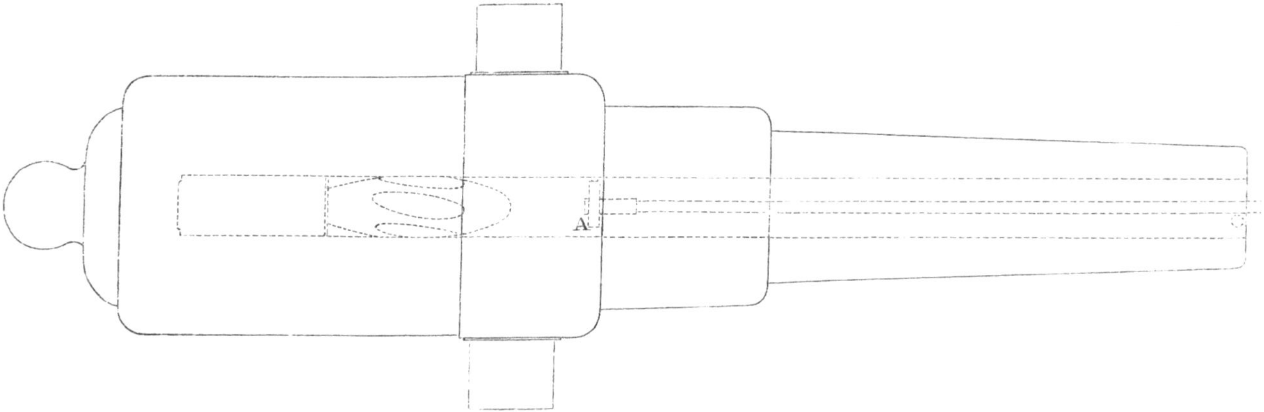

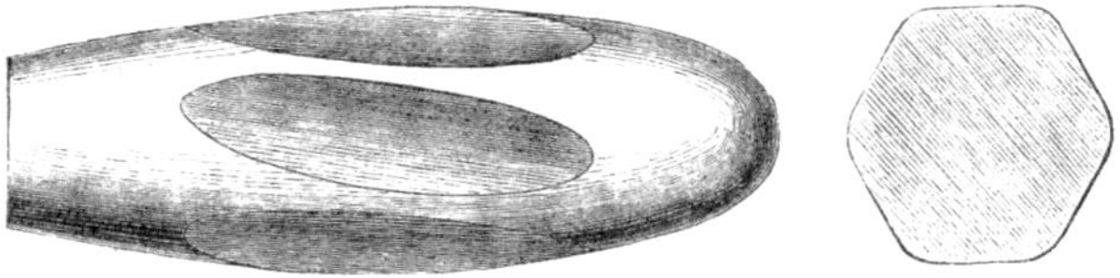

Elevation and Transverse Section of the Projectile ($\frac{1}{8}$).

FIG. 42. 7-in. Gun, with the Measuring Instrument placed in the Bore ($\frac{1}{20}$).

practice. The bore is again measured at the same places, when it will be found that some enlargement has taken place, a result which occurs more or less even with what may be considered a legitimate charge, and is altogether independent of the general overstraining and enlargement produced by an excessive charge.

An increasing weight of shot is now fired with a proper charge of powder until a sensible enlargement not due to wear has been produced. When this occurs the limit of endurance will have been reached, and the greatest charge that can be employed without permanent injury to the gun will have been ascertained.

It is evident that the mode of carrying into effect the principle of the proof by measurement may be varied according to the judgment of the observer. For example, calling the full quantity of powder which it is known that the gun can properly consume the '*proper charge*,' the proof may be commenced by loading the gun with the ordinary weight of shot and with one-half the proper charge of powder. By measurements taken before and after firing the amount of disturbance of the metal, if any, can be determined, and continually increasing charges of powder may be fired in succession until the proper charge is reached. The gun must be measured after each discharge, and there should be no undue or permanent enlargement.

In the trials above referred to measurements were carefully taken at different times during the firing of nearly 3,000 rounds, and the results proved that the enlargements of the bore with successive charges of 10 lbs. of powder and 70-lb. shot were regular, and were due entirely to the wear of the gun in the powder-chamber. In other trials where the powder charge was increased, and a large air space left, the gun being loaded each time with a number of projectiles, the enlargement of the bore became so rapid that a continuation of such charges must have led to the destruction of the gun. In truth, a permanent set took place at every discharge under these conditions.

Unless it be desired to test the ultimate strength of the gun, the firing of two shots of the ordinary weight, with the proper charge of powder, will be a sufficient test that the gun is capable of sustaining its regulation charge of powder and shot.

In a work on 'Guns and Steel,' published in 1873 (see page 66), Sir Joseph Whitworth states:—'Another mode of proof upon the same principle, which I have lately adopted for my guns, consists in preventing the shot from moving when the powder is ignited, the gases generated by the explosion escaping only through the vent.

'I have found that about one-sixth of the regular powder charge fired in this way gives the same strain to the gun as a full charge fired in the ordinary manner. To prevent the movement of the shot a screw is cut on the periphery of the gun, at the muzzle, and on it is fitted a screwed cap having a solid end. The gun is loaded with a cartridge of the ordinary length, but containing one-sixth of the regular charge, and supported by the discs in the centre of the bore; a flat-fronted shot with light wads, to prevent any escape of gas, and a round steel bar, reaching from the shot to the end of the bore, are introduced, and the cap with the solid end is screwed on.

'The gun is then ready for firing, after which my measuring instrument is introduced into the bore, and any enlargement, to the $\frac{10}{1000}$th of an inch in extent, may be ascertained.

'If there be no enlargement, the powder charges may be gradually increased until a slight enlargement has been produced. The real strength of the gun is thus positively ascertained, and should be recorded and stamped upon each gun. This would give confidence to the gunner, and would act as a check upon those engaged in the manufacture.

'When the ultimate endurance of any particular kind of gun is thus to be ascertained, the regular powder charges, or any quantity deemed desirable, may be used, the enlargements being recorded after each discharge.'

The instrument designed for the proof of guns by measurement is shown in figs. 43 and 44. Fig. 43 represents a longitudinal section to a scale of one-fifth, and fig. 44 represents a transverse section of the gun, together with the head of the measuring instrument, on a scale half the full size, and shows the three sliding feelers and contact pieces by which the measurement is effected.

The long stem of the measuring instrument is composed of two tubes, which are made of steel. The outer tube C is attached to a brass head, having three radial arms, D. Three sliding steel feelers, F, are accurately fitted into grooves formed in the respective arms, and are secured in their places by a covering plate. The inner ends of the feelers traverse in three inclined dovetailed grooves in the wedge-piece G, which is attached to the end of the inner tube E; their outer ends, which come in contact with the sides of the bore of the gun, are fitted with pieces of hardened steel, in order to reduce the wear to a minimum. To the outer end of the inner tube E is attached a square threaded screw H, of $\frac{1}{10}$ inch pitch, and to the same end of the outer tube C is attached a brass nut J, working on the screw H. The nut is made in two pieces, which are tightened together endways on the screw, so as to take out backlash or 'lost time.' On the nut is fitted a micrometer-wheel K, having 50 divisions. A slot is cut in the outer tube C, and a pointer I, attached to the inner tube works through the slot and affords the means of measuring accurately the movement of the inner tube.

As the nut J attached to the outer tube is fixed, the inner tube is drawn in or out by $\frac{1}{10}$th inch upon each revolution of the micrometer-wheel; and as the inclined grooves in the wedge-piece G have an inclination of 1 in 20, one division of the micrometer-wheel K causes the feelers F to move outwards or inwards through ·0001 inch or $\frac{1}{10000}$th inch; and an enlargement of the bore to this minute extent of $\frac{1}{10000}$th inch is perceptible upon moving the rod and feelers backwards and forwards a short distance within the bore of the gun. The rod C is supported at the muzzle of the gun on a grooved brass roller L, which facilitates its movement to and fro in the bore of the gun. In order to measure with the greatest accuracy it is necessary that the bore of the gun should be carefully washed out after each discharge, as any fouling or particles of grit left in the bore would interfere with the measurement. A skilful manipulator can readily detect a difference of only one $\frac{1}{10000}$th part of an inch in the dimensions of the bore.

It may here be of interest to give a record of the work done by this measuring machine in some trials made at Shoeburyness in the year 1868. These trials were made by the Ordnance Select Committee, and the object was to compare the ranges obtained from a 9-inch Whitworth muzzle-loading gun, with projectiles weighing 250 lbs. and 310 lbs. respectively.

The small amount of deflection is well worthy of notice, and the range of 11,243 yards has never been obtained by any other gun, whatever its form or calibre.

The details are given in the annexed table:—

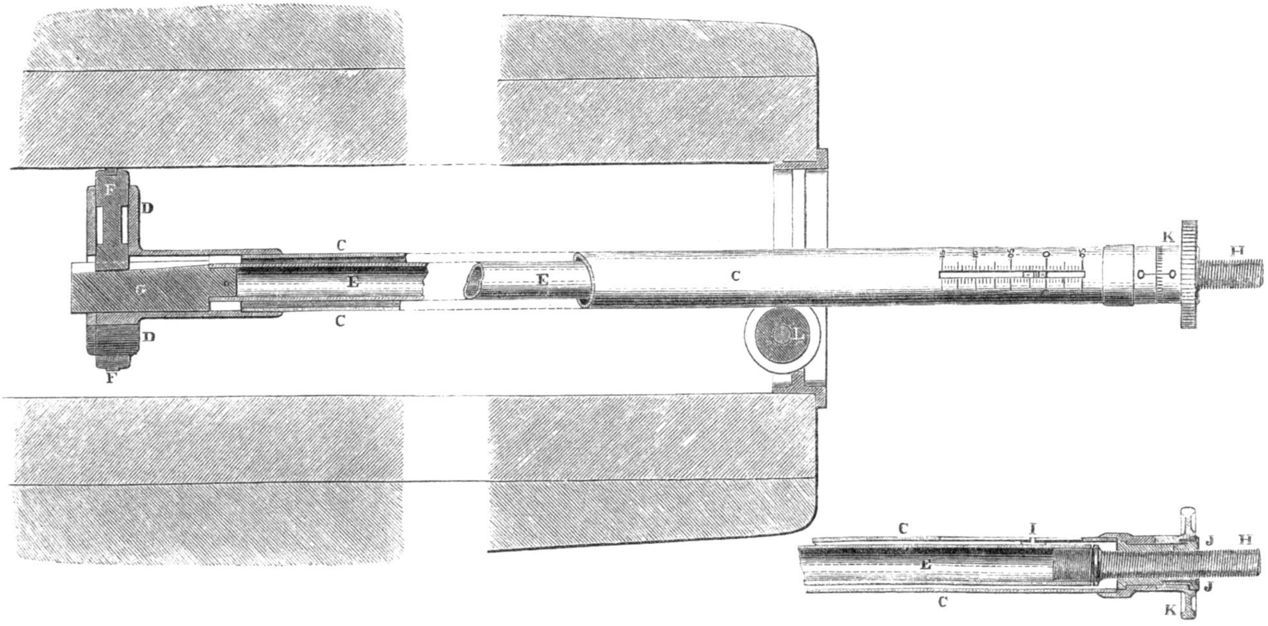

Fig. 43. Details of the Gun-measuring Instrument ($\frac{1}{5}$).

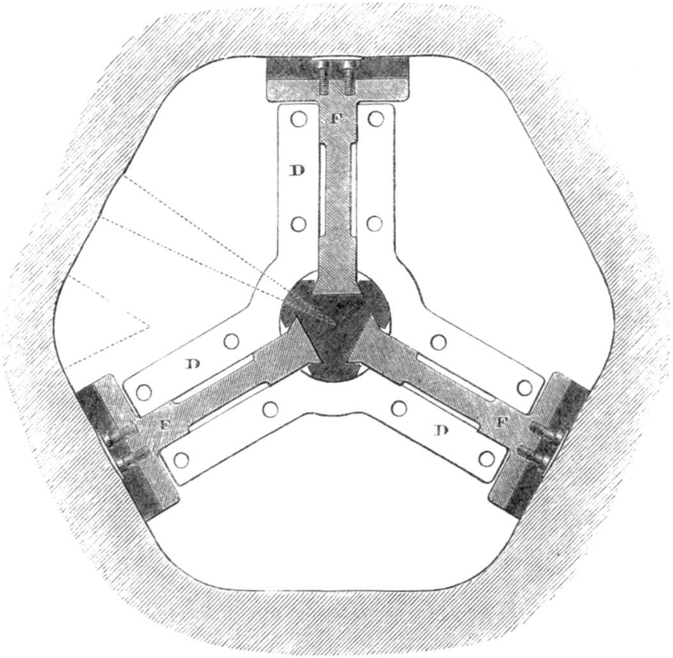

FIG. 44. Transverse Section of the Gun-measuring Instrument ($\frac{1}{2}$).

Date 1868	Round	Eleva-tion	Charge, lbs.	Projectile	Range, yards	Deflection, yards	Wind	Time of flight
Nov. 20	1	10°	50 tubular	310 shell	4,868	1 left	Moderate	13·5 sec.
,,	2	10°	50 ,,	250 shot	5,196	7 right		14·2 ,,
,,	3	10°	50 solid	310 shell	4,021	6⅛ ,,		12·1 ,,
,,	4	33°	50 tubular	250 shot	10,300	3 to 400	Fresh	Not seen
21	5	33°	50 ,,	250 ,,	11,243	yards to		,,
,,	6	33°	50 ,,	310 shell	11,075	the right		,,
,,	7	33° 5′	50 ,,	310 shot	11,127	of line.		,,

The weight of the gun was 14 tons 18 cwt.; the bore was hexagonal, the pitch of the rifling being one turn in 165 inches; the length of the bore was 140·6 inches, the major diameter being 9·05 inches, and the minor diameter 8·20 inches. The length of the 250-lb. shot was 24·5 inches, and that of the 310-lb. shot was 31·6 inches. The powder charge was 50 lbs., the quality of powder being R.L.G.

After these seven rounds had been fired the gun was measured to the $\frac{10}{1000}$th of an inch, and the measurement compared with that before the trial.

In some parts of the bore—viz. at 30 and 32 inches from the breech—there was no enlargement at all; in others, as at distances of 10, 12, 18, 20, 28, 34, and 38 inches, the enlargement varied from $\frac{1}{10000}$th to $\frac{3}{10000}$ths of an inch. At 2 inches from the breech, the enlargement was $\frac{6}{10000}$ths of an inch. At 4, 6, and 8 inches it was $\frac{5}{10000}$ths of an inch, and the greatest enlargement was at 20 and 24 inches, where it amounted to $\frac{7}{10000}$ths of an inch. This was the maximum enlargement after the gun had put forth the power required to throw 250 pounds weight of metal a distance of nearly 6½ miles.

CHAPTER V.

Having now described the measuring machine under its three principal forms, we may appropriately refer to some observations made by Sir Joseph Whitworth upon the character of the work to be accomplished by the use of better apparatus for exact measurement, and by an increased aptitude in applying it. In the 'Miscellaneous Papers on Mechanical Subjects,' published in 1858, we find the following remarks:—

'And here let me state what I think is the proper definition of a good fit. A tight fit is not necessarily a good one; but when the surfaces are true and a proper allowance is made in the size of the parts working together, then a good fit is obtained.

'What constitutes a proper allowance or difference in size depends on the nature of the case and the treatment which the machinery will meet with.

'For instance, the cylinder of the moving headstock of a lathe requires as good a fit as possible, but in practice it is found that the hole in the headstock must be $\frac{1}{2000}$th of an inch larger than the cylinder, and it frequently happens that machinery is not kept in a proper state of cleanliness, or, from motives of false economy, is lubricated with bad oil, in which case there must be a greater difference in size between the parts that have to work together. These are two evils which are productive of great mischief. The abrasion caused by accumulated dust and grit produces increased wear and tear, and soon injures the surfaces in contact;

while bad oil becomes sticky and rancid, and spoils the working of a good fit.

'In machinery supplied to establishments using rape-oil there must be greater allowance and looseness in the fits than would be requisite if better oil, as sperm-oil, were used. I need scarcely say how much more advisable it is to have the more accurate fit, and use the best oil, than to have a loose fit and use the inferior oil, which, causing more friction, consumes greater power.

'A good workman acquires by experience intuitive knowledge of allowances in size which are requisite in various cases, and when he becomes familiar with such sizes as $\frac{1}{1000}$th and $\frac{1}{10000}$th of an inch he will not rest satisfied until he can work with corresponding accuracy. He will also be able to judge of their effect under different circumstances, and know how much to allow in the fitting parts of a machine, according to relative importance and the treatment they are likely to receive at the hands of the attendant.

'As an illustration of the importance of very small differences of size I have made an internal gauge having a cylindrical aperture ·5770 inch diameter, and two external gauges, or solid cylinders, one being ·5769 inch and the other ·5770 inch diameter. The latter is $\frac{1}{10000}$th of an inch larger than the former, and fits lightly in the internal gauge when both are clean and dry; while the smaller, ·5769 inch, gauge is so loose in it as to appear not to fit at all.

'The effect of applying a fine drop of oil to the surfaces of these gauges is very remarkable. It would be observed that the fit of the larger cylinder became more easy, while that of the smaller became more tight. Such results show the necessity of proper lubrication. In the case of the external gauge, ·5770 inch diameter, the external and internal gauges are so near in size that the one does not go through the other when dry, and if pressed in there would be danger of the surface particles of one becoming imbedded in or among those of the other, which I have seen happen, and then no amount of force will separate them; but with a small quantity of oil on their surfaces they move easily and smoothly. In the case of the external gauge, ·5769 inch diameter, which is $\frac{1}{10000}$th of an inch smaller in diameter than the internal gauge, a space of half that quantity is left between the surfaces. This becomes filled with the oil, and hence the tighter fitting which is experienced. It is, therefore, obvious both to the eye and the touch that the difference between these cylinders of $\frac{1}{10000}$th of an inch is an appreciable and important quantity.

'The question of correct measurement is in immediate connection with another which will repay all the attention that can be given to it—and I think that there is no subject that can be more profitably discussed amongst us—I mean that of the *proper gradations of size* in all the various branches of the mechanical arts. I think no estimate can be formed of our national loss from the over-multiplication of sizes. Take, for instance, the various sizes of steam-engines —stationary, marine, and locomotive. In the case of marine engines the number of sizes up to 100-horse power will probably not be short of thirty, when ten perhaps would be ample. If so, look at the sums expended in patterns, designs, and in the number of tools for their manufacture. Nor is this all; for if there were only ten sizes instead of thirty there would be three times the number made of each pattern; and, as you know, the very soul of manufacture is repetition. By attention to this the shipowner would be benefited by getting a better engine at a less price. In the case of locomotives and carriages, for instance, I would urge the subject on the attention of the engineers of the great lines of railway—the London and North-Western, the Midland, the Great Northern, for instance. I hope they will permit me to suggest that they should consider and determine not only the fewest number of sizes of engines and carriages which will suffice, *but also how every single piece may have strictly defined dimensions.* All capricious or uneven dimensions should be avoided; and, as the actual size of the work increases or diminishes, the steps of gradation should be made coarser or finer. This question is also well worthy the attention of our architects and builders. Suppose, for instance, that the principal windows and doors of our houses were made only of three or four different sizes. Then we should have a manufactory start up for making doors without reference to any particular house or builder. They would be kept in stock, and made with the best machinery and contrivances for that particular branch; consequently we should have better doors and windows, at the least possible cost.'

An illustration of the practical value of uniformity of system in mechanical construction, coupled with the adoption of gradations of size, is to be found in the introduction of a uniform system of screw-threads, set forth by Sir Joseph Whitworth at a meeting of the Institution of Civil Engineers, in the year 1841.

The screw-threads which formed the subject of his paper were those of bolts and screws used in fitting up steam-engines and other machinery. The paper stated:—'Great inconvenience is found to arise from the variety of threads adopted by different manufacturers. The general provision for repairs is rendered at once expensive and imperfect. The difficulty of ascertaining the exact pitch of a particular thread, especially when it is not a multiple or submultiple of the common inch measure, occasions extreme embarrassment. This evil would be completely obviated by uniformity of system, the thread becoming constant for a given diameter.'

An extensive collection of screw-bolts having been made from the principal workshops throughout England, the average thread was carefully observed for different diameters. The intervals of $\frac{1}{4}$, $\frac{1}{2}$, 1, and $1\frac{1}{2}$ inches were particularly selected and taken as the fixed points of a scale by which the intermediate sizes were regulated. The only deviation made from the average was such as might be necessary to

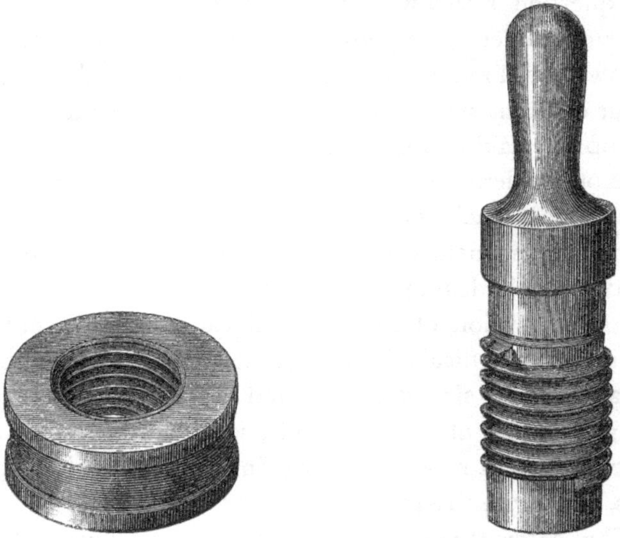

Internal Screw Gauge. External Screw Gauge.

FIG. 45.

avoid the great inconvenience of small fractional parts in the number of threads to the inch.

It is mainly from want of accuracy that screw-bolts so frequently fail. Unless the threads of the screw and nut exactly correspond in every part, and coalesce throughout their whole length and depth, their mutual action is completely deranged, both power and strength are sacrificed, and friction is proportionately increased. The immense consumption of bolts and nuts in fitting up and working machinery may give some idea of the extent to which greater accuracy may be productive of economy.

To maintain uniformity it is necessary that provision should be made for multiplying standards of the diameters and threads. This may easily be done, and will prevent the screwing-tackle from degeneration by use.

In order to preserve uniformity in the threads of screws Sir Joseph Whitworth makes standard external and internal screw-gauges. These gauges are shown in fig. 45. The form of the Whitworth standard thread is shown upon the external screw-gauge. The two cylindrical parts give the exact diameter of the top and bottom of the screw-thread in universal use.

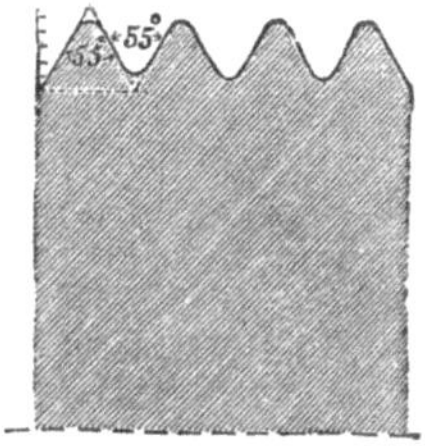

FIG. 46. Diagram of Screw Thread.

The angle made by the opposite sides of the thread is 55° in every case, but the extreme depth which this angle would give is reduced by rounding off the top and bottom, each to the extent of one-sixth, as shown in the diagram (fig. 46). Thus, the depth given to the thread is only two-thirds of that which it would have if the sides intersected, being ·64 of the pitch instead of ·96. A constant proportion is thus established between the depth and pitch of the thread. The precaution of rounding off is adopted to prevent the injury which the thread of the screw and that of the taps and dies might sustain from accident.

The standard pitches which are now universally acknowledged and adopted are given in the table which follows :—

Table of Pitches for Screws with Angular Threads.

No. of Screw Threads per Inch	Old Sizes	New Standards of Size Decimals of an Inch	No. of Screw Threads per Inch	Old Sizes	New Standards of Size Decimals of an Inch	No. of Screw Threads per Inch	Old Sizes	New Standards of Size Decimals of an Inch
48	...	·100	12	...	·525	4½	2⅛	2·125
40	⅛	·125	12	...	·550	4	2¼	2·250
32	...	·150	12	...	·575	4	2⅜	2·375
24	...	·175	12	...	·600	4	2½	2·500
24	...	·200	11	⅝	·625	4	2⅝	2·625
24	...	·225	11	...	·650	3½	2¾	2·750
20	¼	·250	11	...	·675	3½	2⅞	2·875
20	...	·275	11	...	·700	3½	3	3·000
18	...	·300	10	¾	·750			
			10	...	·800	3¼	3¼	3·250
18	...	·325	9	⅞	·875	3¼	3½	3·500
11	...	·350	9	...	·900	3	3¾	3·750
16	⅜	·375	8	1	1·000	3	4	4·000
16	...	·400				2⅞	4¼	4·250
14	...	·425	7	1⅛	1·125	2⅞	4½	4·500
14	...	·450	7	1¼	1·250	2¾	4¾	4·750
14	...	·475	6	1⅜	1·375	2¾	5	5·000
12	½	·500	6	1½	1·500	2⅝	5¼	5·250
			5	1⅝	1·625	2⅝	5½	5·500
			5	1¾	1·750	2½	5¾	5·750
			4½	1⅞	1·875	2½	6	6·000
			4½	2	2·000			

It is important to remark that the proportion between the pitch and the diameter varies throughout the entire scale. Thus, the pitch of ¼ inch is one-fifth of the diameter, that of ½ inch one-sixth, of 1 inch one-eighth, of the 4 inch one-twelfth, of the 6 inch one-fifteenth.

It will be remembered that the threads of which the preceding table shows the average are used in cast iron as well as in wrought iron; and this circumstance has had its effect in rendering them coarser than they would have been if restricted to wrought iron.

The screw which Sir Joseph Whitworth uses in his Works as the standard from which other screws are cut is 30 feet long, with a square thread of ½ inch pitch bounded by a cylinder of 2¾ inches diameter. Every part of the thread from end to end has been carefully tested by an extremely delicate process, in which the measuring machine has borne an important part; and the care and trouble bestowed upon the operation may be judged of when it is stated that Sir Joseph Whitworth devoted a period of six months of his own personal attention and superintendence to making the necessary corrections for perfecting the screw.

It may here be opportune to quote some observations made by Sir Joseph Whitworth on the subdivisions of an inch:—

'There is also another subject which bears upon this question, and which has lately been before the Legislature, viz. that of the decimalising weights and measures. There can be no doubt of the beneficial results that would follow the passing of such a measure. There may be a difference of opinion as to what the unit or integer of lineal measure should be; but I think that it should be the inch, for, from the accuracy with which we can now measure that length, there would be no difficulty in determining and fixing the length of its multiplier. The most important divisions of length in mechanism are those of parts of an inch, and if the length of the inch were altered it would cause much confusion.'

An instructive example occurs in the measurement of thin plates and small wires. Thus, Sir Joseph Whitworth has proposed that the scale for wire gauging should commence with the smallest size and increase by thousandths of an inch, and we quote his observations:—

'Contrary to the custom usually adopted in making the wire gauge, I have called the smallest size No. 1, being $\frac{1}{1000}$th of an inch, No. 2 being $\frac{2}{1000}$ths of an inch, and so on, increasing up to No. 20 by $\frac{1}{1000}$th of an inch between each number; from No. 20 to No. 40 by $\frac{2}{1000}$ths, No. 40 to 100 by $\frac{5}{1000}$ths of an inch. I propose, therefore, to suppress the use of the numbers of designation which have hitherto been employed for the various wire gauges, and simply call the sizes by their expressive numbers in thousandths of an inch, as shown in the accompanying table—a change which will, I think, render the new scale easily intelligible and convenient for use :'—

Table of the Principal Wire and Plate Gauges.

Value of each number in Decimals of an Inch	DECIMAL GAUGE (Thousandths of an inch)	Corresponding Numbers of the Old			Value of each number in Decimals of an Inch	DECIMAL GAUGE (Thousandths of an inch)	Corresponding Numbers of the Old		
		Birmingham Wire Gauge	Birmingham Plate Gauge	Lancashire Gauge			Birmingham Wire Gauge	Birmingham Plate Gauge	Lancashire Gauge
·001	1	...	...	...	·045	45	...	15*	56
·002	2	...	...	...	·050	50	18	16	55
·003	3	...	...	...	·055	55	...	17*	54
·004	4	36	1	...	·060	60	17*	18	52
·001	5	35	2	...	·065	65	16	19	51
·006	6	...	...	...	·070	70	15*	21*	49
·007	7	34	...	...	·075	75	...	22	47
·008	8	33	3	...	·080	80	...	24*	45
·009	9	32	...	...	·085	85	14*	...	43
·010	10	31	4	...	·090	90	...	...	42
·011	11	...	...	...	·095	95	13	25	41
·012	12	30	5	...	·100	100	...	26**	38
·013	13	29	6	80	·110	110	12	27**	34
·014	14	28	...	79	·120	120	11	28	31*
·015	15	...	7	78	·135	135	10	31*	29
·016	16	27	8	77	·150	150	9*	34*	23
·017	17	...	...	...	·165	165	8	36*	19
·018	18	26	...	76	·180	180	7	...	13
·019	19	...	9	75	·200	200	6**	...	5
·020	20	25	...	...	·220	220	5	...	2
·022	22	24	...	74	·240	240	4*	...	C
·024	24	23	10	72	·260	260	3	...	G
·026	26	...	...	71	·280	280	2**	...	K
·028	28	22	11	70	·300	300	1	...	N*
·030	30	...	...	68	·325	325	...	...	P*
·032	32	21	...	66	·350	350	...	...	S*
·034	34	...	12	64	·375	375	00**	...	V*
·036	36	20	13	62	·400	400	...	...	X**
·038	38	...	...	61	·425	425	000	...	...
·040	40	19	14	59	·453	450	0000**	...	...
					·475	475			
					·500	500			

NOTE—Sizes which differ from those in the first column by more than ·002 of an inch are marked thus **; those of which the difference exceeds ·001, thus *. All others either correspond exactly or are within ·001 of an inch.

In the Whitworth Decimal Gauge, shown in fig. 47, the notches are made by drilling a series of holes some little distance from the edges of the plate, with which they are then connected by saw-cuts of varying widths, the sides of the cuts being afterwards made smooth and parallel by driving into each notch a hard steel drift of the exact thick-

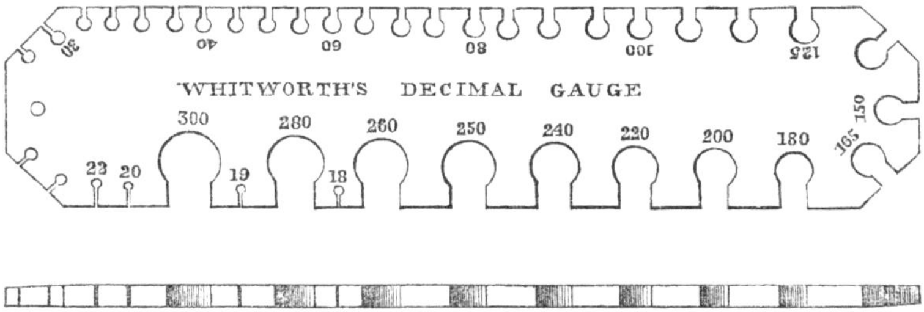

FIG. 47 ($\frac{2}{3}$).

ness represented by the number with which the notch is to be marked. The decimal wire gauges to which the above table refers are made from standard flat gauges, one of which is shown in fig. 48. The two faces of the gauge are made true parallel planes. A separate gauge of this form is made for each notch in the Decimal Wire Gauge from

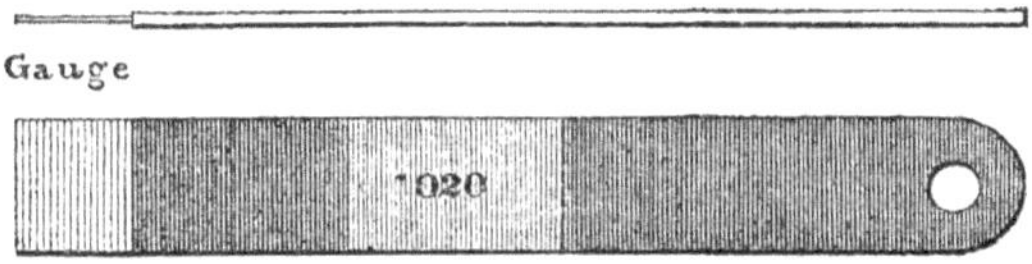

FIG. 48. Flat Gauge (full size).

No. 20 to No. 95 inclusive; below No. 20 the Workshop Measuring Machine must be used for adjusting the gauge, and for sizes above No. 95 cylindrical gauges are employed. These standard gauges are put together in sets, and serve not only for the manufacture but also for the correction of the notched gauge.

After the lapse of many years of labour, and after striving in vain to induce the authorities of the Standard Department to adopt, for the benefit of the public, that system of end measurement which appears to be of the highest practical value, it will be found that the proposals originally pressed upon the Committee in the year 1855 have been adopted in a modified form. In 1855, Sir Joseph Whitworth, in giving his evidence before the Standards Committee, stated :—

'One of the most important things which could be done with reference to standard measures would be to give to each manufacturing town a complete set of standards of end measure. . . . What is required is a standard of two feet in length, a standard foot, a standard inch, and standards below that; for instance, a standard of $\frac{1}{100}$th of an inch, with differences to hundredths of an inch.

'Since I discovered this exact mode of measuring which I have explained, we have made standards in great numbers, and we have had many applications to supply the measuring apparatus; but, as the public had confidence in me in making these standards, I thought it was not desirable to do so, and have accordingly refused. But I consider that the responsibility of furnishing correct standards ought not to rest with me, but with the Government, and I am prepared to give my measuring machine to the Government on any terms which they may think proper; and thus the public generally might be furnished with these machines, which would be of great advantage to them.'

The offer so made was not accepted. The Standards Committee had decided that all primary standards in end-measure bars should have rounded ends, and that secondary standards only should comprise end measure bars with plane ends. But inasmuch as we have shown that the accuracy of comparison of end-measure bars with rounded ends depends solely on line measure, this reservation was equivalent to shutting out the primary standards from the use both of engineers and of the general public. It has been abundantly shown that the only comparison available in the workshop is that of end measure between plane surfaces.

Sir Joseph Whitworth has always contended that available standards of length, in the form of end-measure bars with plane ends, should be placed in the care of corporate bodies in large cities, and should be accessible, under proper restrictions, for the use of the public. He has contended that the inch should be made the standard of length, and that it should be subdivided decimally.

The advantage accruing to the public by the adoption of accessible standards for end-measures of length was thus suggested to the Standards Department, but the Board of Works did not take action until the spring of the present year (1876), when they, for the first time, erected mural standard measures on the north side of Trafalgar Square. These standards are formed on a bronze plate securely fastened by screws to a solid frame which is embedded in a sloping granite wall. The end measures are of 1, 2, and 3 feet, given by parallel plane surfaces, and there are nine subdivisions of these measures, including those of an inch divided into tenths. The standards will only give a correct indication at a temperature of 62° Fahr.

The result to be particularly noticed is that the inch has been subdivided decimally, and has virtually become the unit of measurement.

The nature of Sir Joseph's measuring instruments has now been fully explained, the delicacy of the observation and the limits of refined accuracy to which it can be carried has been pointed out, and it may well be asked whether a primitive appliance set up in the open air, in positions where it can only be accurate at one definite temperature, is of much use? The thing to be done is of easy accomplishment, and may be done well under proper conditions, but it is surely not enough to present a bronze standard, with a surface affected by exposure to the atmosphere, in a position which negatives the possibility of an exact observation, and to rely on that as a fulfilment of the wants of mechanicians throughout this country. What is required is accessible measuring apparatus which will permit of the comparison of new gauges with those of a standard size—not by the rude method in Trafalgar Square, but by those practicable and reliable instruments which will indicate to the $\frac{1}{10000}$th or the $\frac{1}{100000}$th of an inch, or even to more narrow limits.

We have now given the details of one great work upon which Sir Joseph Whitworth has expended a vast amount of labour; and in thus pressing his views as to the importance of correct measurement, and as to the advantages to be anticipated from the general distribution of reliable standards of end measure, we conclude this notice of his Measuring Machines.

LONDON: PRINTED BY
SPOTTISWOODE AND CO., NEW-STREET SQUARE
AND PARLIAMENT STREET

Made in United States
Troutdale, OR
03/27/2026

48059935R00066